PHYSICAL METHODS OF CHEMISTRY

Second Edition

Editors: Bryant W. Rossiter
           Roger C. Baetzold

PHYSICAL METHODS OF CHEMISTRY

Second Edition

Edited by

BRYANT W. ROSSITER

and

ROGER C. BAETZOLD

Research Laboratories
Eastman Kodak Company
Rochester, New York

Volume VII

DETERMINATION OF ELASTIC AND MECHANICAL PROPERTIES

A WILEY-INTERSCIENCE PUBLICATION

JOHN WILEY & SONS, INC.

New York · Chichester · Brisbane · Toronto · Singapore

Library of Congress Cataloging-in-Publication Data:

Determination of elastic and mechanical properties / edited by Bryant
    W. Rossiter and Roger C. Baetzold.
        p.   cm.—(Physical methods of chemistry. Second edition,
    ISSN 0162-7627 ; v. 7)
    "A Wiley-Interscience publication."
    Includes bibliographical references and index.
    ISBN 0-471-53438-2
    1. Materials--Testing.   2. Elasticity.   I. Rossiter, Bryant W.,
    1931–   .   II. Baetzold, Roger C.   III. Series: Physical methods of
    chemistry (2nd ed.) ; v. 7.
    [TA410]
    542 s—dc20
    [620. 1'1292]   90-13009
                                                        CIP

Printed in the United States of America

10 9 8 7 6 5 4 3 2 1

CONTRIBUTORS

GARRON P. ANDERSON,† Morton Thiokol, Inc., Brigham City, Utah

HUGH R. BROWN, IBM Almaden Research Center, San Jose, California

LEIF A. CARLSSON, Department of Mechanical Engineering, Florida Atlantic University, Boca Raton, Florida

GREGORY D. DEAN, Division of Materials Metrology, National Physical Laboratory, Teddington, Middlesex, England

K. LAWRENCE DEVRIES, College of Engineering, University of Utah, Salt Lake City, Utah

JOHN C. DUNCAN, Division of Materials Metrology, National Physical Laboratory, Teddington, Middlesex, England

NORMAN S. EISS JR., Department of Mechanical Engineering, Virginia Polytechnic Institute and State University, Blacksburg, Virginia

BRUCE HARTMANN, Naval Surface Warfare Center, Silver Spring, Maryland

CAROL R. JOHNSEN, College of Engineering, University of Utah, Salt Lake City, Utah

BRYAN E. READ, Division of Materials Metrology, National Physical Laboratory, Teddington, Middlesex, England

DALE W. WILSON, Applications and Product Development, CELION Carbon Fibers, A Subsidiary of BASF Structural Materials, Inc., Charlotte, North Carolina

†Deceased, October, 1988.

PREFACE TO PHYSICAL METHODS OF CHEMISTRY

This is a continuation of a series of books started by Dr. Arnold Weissberger in 1945 entitled *Physical Methods of Organic Chemistry*. These books were part of a broader series, *Techniques of Organic Chemistry*, and were designated Volume I of that series. In 1970, *Techniques of Chemistry* became the successor to and the continuation of the *Techniques of Organic Chemistry* series and its companion, *Techniques of Inorganic Chemistry*, reflecting the fact that many of the methods are employed in all branches of chemical sciences and the division into organic and inorganic chemistry had become increasingly artificial. Accordingly, the fourth edition of the series, entitled *Physical Methods of Organic Chemistry*, became *Physical Methods of Chemistry*, Volume I in the new *Techniques* series. The last edition of *Physical Methods of Chemistry* has had wide acceptance, and it is found in most major technical libraries throughout the world. This new edition of *Physical Methods of Chemistry* will consist of eight or more volumes and is being published as a self-standing series to reflect its growing importance to chemists worldwide. This series will be designated as the second edition (the first edition, Weissberger and Rossiter, 1970) and will no longer be subsumed within *Techniques of Chemistry*.

This edition heralds profound changes in both the perception and practice of chemistry. The discernible distinctions between chemistry and other related disciplines have continued to shift and blur. Thus, for example, we see changes in response to the needs for chemical understanding in the life sciences. On the other hand, there are areas in which a decade or so ago only a handful of physicists struggled to gain a modicum of understanding but which now are standard tools of chemical research. The advice of many respected colleagues has been invaluable in adjusting the contents of the series to accommodate such changes.

Another significant change is attributable to the explosive rise of computers, integrated electronics, and other "smart" instrumentation. The result is the widespread commercial automation of many chemical methods previously learned with care and practiced laboriously. Faced with this situation, the task of a scientist writing about an experimental method is not straightforward.

Those contributing to *Physical Methods of Chemistry* were urged to adopt as their principal audience intelligent scientists, technically trained but perhaps inexperienced in the topic to be discussed. Such readers would like an introduction to the field together with sufficient information to give a clear understanding of the basic theory and apparatus involved and the appreciation for the value, potential, and limitations of the respective technique.

Frequently, this information is best conveyed by examples of application, and many appear in the series. Except for illustration, however, no attempt is made to offer comprehensive results. Authors have been encouraged to provide ample bibliographies for those who need a more extensive catalog of *applications*, as well as for those whose goal is to become more expert in a *method*. This philosophy has also governed the balance of subjects treated with emphasis on the *method*, not on the results.

Given the space limitations of a series such as this, these guidelines have inevitably resulted in some variance of the detail with which the individual techniques are treated. Indeed, it should be so, depending on the maturity of a technique, its possible variants, the degree to which it has been automated, the complexity of the interpretation, and other such considerations. The contributors, themselves expert in their fields, have exercised their judgment in this regard.

Certain basic principles and techniques have obvious commonality to many specialties. To avoid undue repetition, these have been collected in Volume I. They are useful on their own and serve as reference material for other chapters.

We are deeply sorrowed by the death of our friend and associate, Dr. Arnold Weissberger, whose enduring support and rich inspiration motivated this worthy endeavor through four decades and several editions of publication.

BRYANT W. ROSSITER
JOHN F. HAMILTON

Research Laboratories
Eastman Kodak Company
Rochester, New York
March 1986

PREFACE

The subject matter of this volume is entirely new to the *Physical Methods of Chemistry* series as it was not treated in the earlier editions. While determination of elastic and mechanical properties has always been important to some industrial laboratories, the significance of these measurements has increased tremendously in recent years for academic and industrial scientists and engineers as a result of new advanced materials research and highly sophisticated, automated manufacturing and processing methods. Materials such as metals, alloys, oxides, polymers, and composites possess unique chemical and physical properties, many of which are peculiar to the high technology age and to where these properties must be measured, maintained, controlled, and understood under the most exacting and demanding conditions. The chapters in this volume are written by world-class authors who are widely recognized in their fields and who have broad practical laboratory experience in the application of their respective techniques. The book is adapted to the needs of intelligent nonexperts who desire to make measurements for the first time or to the more seasoned experimenter who wishes to apply the most recent methods to new discovery, quality control, or manufacturing purposes.

This volume is more than a review of the various subjects; rather, contributors provide, either directly or through clearly designated references, information that is essential to the use of the techniques in the laboratory. The recent rapid advances in materials sciences have made this volume an essential and indispensable part of a modern technical library.

We acknowledge our deep gratitude to the contributors who have spent long hours over manuscripts, and we welcome to Volume VII and to the series: Dr. Hugh R. Brown, Professor Leif A. Carlsson, Dr. Gregory D. Dean, Professor K. Lawrence DeVries, Dr. John C. Duncan, Professor Norman S. Eiss Jr., Dr. Bruce Hartmann, Ms. Carol R. Johnsen, Dr. Bryan E. Read, and Mr. Dale W. Wilson. As well, we extend our sympathy and gratitude to the colleagues and family of Dr. Garron P. Anderson.

We are also extremely grateful to the many colleagues from whom we have sought counsel on the choice of subject matter and contributors. We express our gratitude to Mrs. Ann Nasella for her enthusiastic and skillful editorial assistance. In addition, we heartily thank the specialists whose critical readings of the manuscripts have frequently resulted in the improvements accrued from collective wisdom. For Volume VII they are Dr. R. A. Bubeck, Mr. J. R. Dann, Dr. D. J. Massa, Dr. J. E. Masters Jr., Dr. J. M. O'Reilly, Dr. M. B. Peterson, Dr. E. Plueddemann, and Professor E. Rabinowicz.

BRYANT W. ROSSITER
ROGER C. BAETZOLD

Rochester, New York
January 1991

CONTENTS

†Deceased, October, 1988.

PHYSICAL METHODS OF CHEMISTRY
Second Edition

Volume VII

DETERMINATION OF ELASTIC AND MECHANICAL PROPERTIES

Chapter **1**

DETERMINATION OF DYNAMIC MODULI AND LOSS FACTORS

Bryan E. Read, Gregory D. Dean, and John C. Duncan

1   INTRODUCTION

Dynamic mechanical techniques are used increasingly to characterize the structures and relaxation behavior of viscoelastic materials and to obtain materials property data for a range of applications. With these methods the deformation of samples subjected to oscillating forces are measured, and dynamic moduli and loss factors are evaluated as a function of forcing frequency and temperature. Details of such techniques appear in numerous original papers and in some general reviews and books [1–9].

This chapter presents a broad survey of dynamic methods developed for the study of solid materials at frequencies between 10^{-2} and 10^7 Hz. These include low-frequency, nonresonance methods (Section 2), the torsion pendulum (Section 3), audiofrequency resonance techniques (Section 4), and an ultrasonic method (Section 5) involving measurements of the velocity and attenuation of the component waves of pulses transmitted through the material. Since details of most methods can be found in the original references, emphasis is given to the principles, range, and accuracy of the techniques with the aim of demonstrating the reliability of data for different materials over wide frequency and temperature ranges. Section 1.4 also gives a brief account of the relevance of dynamic data to various engineering applications and to studies of mechanical relaxation and associated molecular dynamics. Illustrative new data are subsequently presented for an immiscible polymer blend, for poly(propylene), and for anisotropic carbon fiber reinforced poly(ether ether)ketone (PEEK). Methods developed for the study of liquids [3, 10–13] are mentioned only briefly. These are often similar in principle to the techniques employed for solids, and they were discussed in the previous edition of this series [14].

1.1   Moduli and Loss Factors for Isotropic Materials

In discussing the quantities used to specify the dynamic properties of viscoelastic materials [7, 15] we first consider an isotropic sample that is subjected to a sinusoidally varying tensile strain ϵ at a frequency below that required to induce resonance vibrations. For linear viscoelastic behavior, under steady-state conditions, the stress σ sustained by the sample is also sinusoidal; but the stress

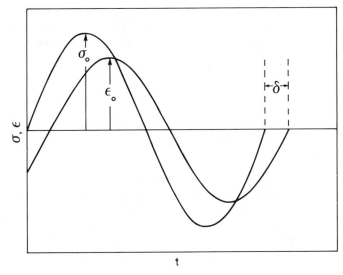

Figure 1.1 Time dependence t of the stress σ and strain ϵ during a low-frequency (nonresonance) test.

cycle leads the strain cycle by some phase angle δ, as illustrated in Figure 1.1. We thus write

$$\epsilon = \epsilon_0 \cos \omega t \tag{1}$$

$$\sigma = \sigma_0 \cos(\omega t + \delta)$$
$$= \sigma_0 \cos \delta \cos \omega t - \sigma_0 \sin \delta \sin \omega t \tag{2}$$

where ω is the angular frequency and σ_0 and ϵ_0 are the amplitudes of σ and ϵ. Equation (2) demonstrates that the stress consists of two components. One component is of magnitude $\sigma_0 \cos \delta$ and it is in phase with the strain. The other component of magnitude $\sigma_0 \sin \delta$ is 90° ahead of the strain and thus in phase with the strain rate. Therefore, the material behaves partly as an elastic solid and partly as a viscous liquid, and the stress–strain relationship is written

$$\sigma = \epsilon_0 E' \cos \omega t - \epsilon_0 E'' \sin \omega t \tag{3}$$

where the component moduli are given by

$$E' = \frac{\sigma_0}{\epsilon_0} \cos \delta \quad \text{and}$$

$$E'' = \frac{\sigma_0}{\epsilon_0} \sin \delta \tag{4}$$

These equations suggest that the tensile modulus can be specified in complex form. For this purpose the strain and stress cycles are represented by the real parts of $\epsilon^* = \epsilon_0 \exp(i\omega t)$ and $\sigma^* = \sigma_0 \exp[i(\omega t + \delta)]$, respectively, where $i = (-1)^{1/2}$. Then

$$E^* = \frac{\sigma^*}{\epsilon^*} = \frac{\sigma_0}{\epsilon_0} \exp(i\delta)$$

$$= \frac{\sigma_0}{\epsilon_0} (\cos \delta + i \sin \delta) = E' + iE'' \tag{5}$$

The real part of the modulus E', which is in phase with the strain, is termed the *storage modulus* since it is proportional to the peak energy stored per cycle in the material. The imaginary part of the modulus E'', which is out of phase with the strain, is proportional to the net energy dissipated per cycle and is known as the *loss modulus*. The ratio

$$\frac{E''}{E'} = \tan \delta_E \tag{6}$$

is termed the *loss factor* or *damping factor*, and the subscript E is added to δ to indicate that the phase angle is measured in a tensile test. It should be emphasized that E', E'', and $\tan \delta_E$ depend on the test frequency and also on temperature, and each is used to characterize dynamic mechanical properties either at a given frequency or temperature or, preferably, over a range of these variables.

During the dynamic tensile tests on solid strips, measurements may also be made of the ratio of amplitudes v_0 and phase angle δ_v between the lateral and longitudinal strain cycles. These measurements yield the components of a complex Poisson's ratio [7, 15]

$$v^* = v' - iv''$$

where

$$v' = v_0 \cos \delta_v$$

$$v'' = v_0 \sin \delta_v$$

$$\frac{v''}{v'} = \tan \delta_v \tag{7}$$

and positive values of v'' and δ_v signify that the lateral strain lags in phase behind the longitudinal strain.

Although the components of E^* are determined quite simply from dynamic tensile or flexural tests, it is often convenient or desirable to obtain dynamic properties for other modes of deformation. The two basic types of deformation correspond to shear, which involves a change of shape at constant volume, and to isotropic compression, where a volume change occurs at constant shape. For isotropic materials, dynamic experiments employing either simple shear or torsional deformations yield the components of the complex shear modulus

$$G^* = G' + iG'' = G'(1 + i \tan \delta_G) \tag{8}$$

where G', G'', and $\tan \delta_G$ are the shear storage modulus, loss modulus, and loss factor, respectively, while measurement of the volume response to alternating hydrostatic pressure gives the components of the complex bulk modulus

$$K^* = K' + iK'' = K'(1 + i \tan \delta_K) \tag{9}$$

where K', K'', and $\tan \delta_K$ are the bulk storage modulus, loss modulus, and loss factor, respectively.

From dynamic tests where alternating normal forces are applied to the faces of a wide, thin sheet such that the lateral strains are negligible, the components of a complex longitudinal modulus

$$L^* = L' + iL'' = L'(1 + i \tan \delta_L) \tag{10}$$

are obtained. Here L', L'', and $\tan \delta_L$ are the longitudinal storage modulus, loss modulus, and loss factor, respectively. These components are relevant to the propagation of longitudinal ultrasonic waves in liquids or solids (see Section 5) when elements of the material cannot move laterally to a significant extent.

Each preceding component modulus and loss factor depends on frequency and temperature and has a significance analogous to that discussed for the corresponding tensile property. Furthermore, the dynamic properties characteristic of the different types of deformation are not independent. For isotropic materials, measurements of storage moduli and loss factors as a function of frequency for two deformation modes are sufficient to characterize the dynamic viscoelastic properties completely. Relationships between the component moduli and loss factors can be derived from the equations of classical elasticity theory expressed in complex forms [7, 15]. For example,

$$G^* = E^*/2(1 + v^*) \tag{11}$$

$$K^* = E^*/3(1 - 2v^*) \tag{12}$$

$$K^* = L^* - (4G^*/3) \tag{13}$$

$$E^* = G^*(3L^* - 4G^*)/(L^* - G^*) \tag{14}$$

After separating the real and imaginary components in (11) and (12), we can obtain the components of G^* and K^*, often with small error magnifications [9, 16], from the measured components of E^* and v^* in low-frequency tensile tests. Similarly (13) and (14) allow us to determine the components of K^* and E^* from measurements of L^* and G^* components at ultrasonic frequencies [7].

1.2   Complex Moduli for Anisotropic Materials

Many solids are anisotropic, and the storage moduli and loss factors depend on the direction of loading [7]. Anisotropy in crystalline solids, for example, arises from the packing geometry of atoms or molecules in the crystal, and for natural polymers (e.g., wood) or synthetic polymers it may originate from the preferential orientation of chain molecules or reinforcing fibers. The number of independent complex moduli required to characterize fully the dynamic properties of an anisotropic material depends on its structural symmetry. For materials with orthorhombic symmetry this number is 9, whereas for hexagonally symmetric materials, the number is 5. In the latter, which includes uniaxially drawn polymers or unidirectionally reinforced composites, we define Cartesian axes 1, 2, and 3 within the specimen where direction 1 is parallel to the unique symmetry axis and directions 2 and 3 correspond to two equivalent orthogonal directions. Then the independent moduli can be E_1^*, E_2^*, G_{12}^*, G_{23}^*, and v_{12}^* [7]. Here E_1^* and E_2^* are the longitudinal and transverse moduli measured in directions 1 and 2, respectively; G_{12}^* and G_{23}^* are shear moduli corresponding to shear stresses in directions 2 and 3 and on planes perpendicular to 1 and 2, respectively; and v_{12}^* $(=\epsilon_2^*/\epsilon_1^*)$ is the complex Poisson's ratio for loading in direction 1.

1.3   Complex Compliances and Relationships to Creep Compliances

The dynamic properties of linear viscoelastic materials can also be represented by the components of a complex compliance. For example, in dynamic tensile experiments on an isotropic material we define a complex tensile compliance

$$D^* = D' - iD'' \tag{15}$$

where D' and D'' are the tensile storage and loss compliances, respectively. Since $D^* = 1/E^*$,

$$D' = \frac{E'}{E'^2 + E''^2}$$

$$D'' = \frac{E''}{E'^2 + E''^2}$$

$$\frac{D''}{D'} = \frac{E''}{E'} = \tan \delta_E \tag{16}$$

Complex compliances are similarly defined for other modes of deformation such as shear ($J^* = 1/G^*$) or hydrostatic compression ($B^* = 1/K^*$) as well as for anisotropic materials [7]. In addition the compliance components also depend on frequency and temperature, but they are less frequently employed than the modulus components for the characterization of dynamic behavior.

Provided that the viscoelastic behavior is linear, the components of the complex moduli or compliances can be related to quantities derived from other viscoelastic measurements such as stress–relaxation moduli or creep compliances [15]. As an example, we note that for $\tan \delta_E \leqslant 0.1$, the tensile creep compliance $D(t)$ is given to within about 1% accuracy by [17],

$$D(t) = D'(1/t) + 0.337D''(0.323/t) \qquad (17)$$

where D' and D'' are measured at the angular frequencies $1/t$ and $0.323/t$ as indicated. A simpler approximation

$$D(t) = D'(\kappa/t) \qquad (18)$$

where κ approximately equals 0.6, was found [18] to yield an accurate transformation of dynamic data for poly(methyl methacrylate) at room temperature. Similar approximations to (17) and (18) apply to results from other types of deformation.

1.4   Data Applications

The dynamic properties of materials in different frequency ranges are directly relevant to the design of engineering components when stiffness and damping capacity are important. Relevant applications include antivibration mounts, the sound-deadening property of panels, and acoustic and sonar devices [7, 19–22]. In addition, the results of dynamic studies contributed much to our understanding of the motions of atoms or molecules, which have an important bearing on a wide range of materials properties. A brief outline of mechanical relaxation effects in relation to materials structure indicates the importance of dynamic measurements in this context.

Figure 1.2 illustrates schematically the variations of E', E'', D', D'', and $\tan \delta_E$ with $\log \omega$ at constant temperature. In a relaxation region E' and D' increase and decrease, respectively, with increasing frequency between limiting relaxed E_R, D_R and unrelaxed E_U, D_U values. Maxima occur in the E'' and D'' curves at frequencies close to the inflection points in the respective E' and D' plots, and a peak in $\tan \delta_E$ is observed at some intermediate frequency.

The rates of atomic or molecular rearrangements in metals [1] and simple liquids [23] are frequently characterized by a single relaxation time τ determined from the frequency $\omega_{max} = 1/\tau$ at which either E'', $\tan \delta_E$, or D'' exhibits a maximum. We note that the different time constants τ_E, τ_δ, and τ_D, which are indicated in Figure 1.2, are related by the linear theory of viscoelasticity [5]. Furthermore, for processes characterized by a single relaxation time the loss

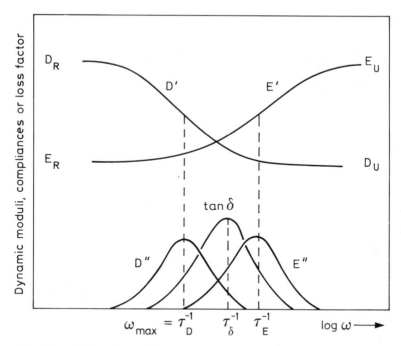

Figure 1.2 Schematic dependence of the complex modulus and compliance components on angular frequency ω.

peaks are quite sharp, having a width of 1.14 decades at a loss level equal to one-half the peak magnitude.

According to the barrier models of relaxation, the motions of atoms or molecules involve transitions over a potential energy barrier between equilibrium states of free energy F_1 and F_2 when the equilibrium distribution of relaxing entities between the two states is perturbed by the applied stress [5] or by cyclic temperature variations in compressional waves [23]. The relaxation time is related to the barrier height or activation energy $\Delta F\ddagger$ by an Arrhenius or Eyring equation

$$\tau = \tau_0 \exp(\Delta F\ddagger / RT) \qquad (19)$$

where R is the gas constant, T is the absolute temperature, and τ_0 is the reciprocal frequency of atomic or molecular vibrations. Values of $\Delta F\ddagger$ are, therefore, derived from the temperature dependence of ω_{max}. The relaxation strength $E_U - E_R$ is predicted to depend on the density of relaxing entities and on $(F_2 - F_1)/RT$ [1, 23]. For various molecules in the liquid state conformational energies $F_2 - F_1$ were evaluated from the temperature dependence of ultrasonic absorption magnitudes [23].

For polymers [5] the observed loss peaks are generally much broader than that for a single relaxation time, and several techniques may be required to obtain results covering a complete relaxation region. Such data are conveniently interpreted by empirical distributions of relaxation times, and values of $1/\omega_{max}$ serve to define some average relaxation time $\bar{\tau}$. Because of limitations in available frequency ranges, we usually obtain dynamic data as a function of temperature at constant frequency. Loss peaks are then observed at temperatures for which $1/\bar{\tau}$ equals the test frequency.

In Figure 1.3a four regions of mechanical behavior are schematically illustrated for an uncross-linked amorphous polymer. These correspond to a glassy β region, a glass–rubber α relaxation region, a rubbery plateau, and a flow region. Secondary relaxations in the glassy region are usually ascribed to local motions of polymer chain segments or of side groups attached to the main chain, and the temperature dependence of $\bar{\tau}$ for these processes is usually described by the Arrhenius equation (19).

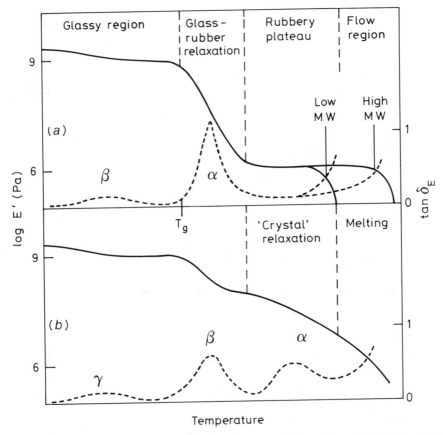

Figure 1.3 Typical variation of the storage modulus and loss factor with temperature at constant frequency for (a) an uncross-linked amorphous polymer and (b) a partially crystalline polymer.

The onset of long-range micro-Brownian motions occurs at the glass–rubber relaxation region and gives rise to a large decrease in modulus and pronounced loss peak. At frequencies around 1 Hz this peak occurs about 20°C above the glass transition temperature T_g as determined by the incremental change in thermal expansion coefficient or heat capacity [24]. For the glass–rubber relaxation the temperature dependence of $\bar{\tau}$ is non-Arrhenius and consistent with the empirical Williams, Landel, and Ferry (WLF) equation or equivalent Vogel–Fulcher relationship

$$\bar{\tau} = \tau_0 \exp[A/(T - T_\infty)] \tag{20}$$

Here τ_0 and A are constants and T_∞ (ca. $T_g - 50°C$) is the temperature at which $\bar{\tau}$ becomes infinite at structural equilibrium. In practice nonequilibrium structures are produced in the material on cooling to temperatures below T_g. Measured values of $\bar{\tau}$ are then less than the equilibrium values, and the glass–rubber relaxation is shifted to higher frequencies or lower temperatures. For an amorphous polymer held at some constant temperature below T_g, structural changes give rise to a gradual increase in $\bar{\tau}$ toward its equilibrium value. These physical aging effects [25] have an important influence on the long-term creep and other mechanical properties of polymers, and they were recently investigated with the aid of dynamic techniques [18, 26, 27].

In the flow region the slippage of chain molecules relaxes the stress sustained by entanglements in the rubbery plateau. The location of the flow region depends strongly on molecular weight, as indicated in Figure 1.3a, and the terminal decrease in modulus is accompanied by a continuous rise in loss factor. For partially crystalline polymers chain slippage is prevented by the cross-linking effect of the crystals, and the flow region is eliminated below the melting point. However, the presence of crystals yields a relaxation labeled α (Figure 1.3b) attributed [28] to chain motions within the crystal that couple with rearrangements of restrained disordered material at the crystal surfaces. The glass–rubber relaxation for partially crystalline polymers, associated with motions within the unrestrained amorphous phase, is now labeled β.

2   FORCED VIBRATION NONRESONANCE METHOD

At low frequencies (typically 0.01–100 Hz) dynamic moduli and loss factors are determined directly from the amplitudes of, and phase angle between, the force and displacement cycles for samples subjected to a time-harmonic force or deformation. Although this nonresonance technique is employed for studies of various materials ranging from viscoelastic liquids [3, 10] to rigid solids [3–7], this section is concerned predominantly with applications of the method to solids, including rubbers. Commercial instruments within this category, some of which are listed in Table 1.1, have contributed much to the automatic acquisition of dynamic data over wide temperature ranges.

2.1 Apparatus and Basic Theory

Figures 1.4 and 1.5 illustrate, respectively, two types of nonresonance instrument that are extensively employed in the study of solids. With the equipment shown in Figure 1.4, harmonic displacements are generated by a vibrator V and monitored by a displacement transducer D. Various modes of sinusoidal deformation are then induced in samples S through suitable design of loading stage and specimen geometry, and a force transducer F is employed to measure the restoring forces in the sample. Such instruments include the Rheovibron

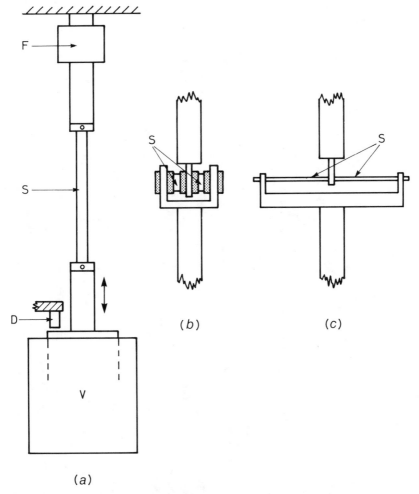

(a)

Figure 1.4 Illustration of the nonresonance method employing a force transducer F and loading stages for (a) tensile, (b) simple shear, and (c) clamped–clamped flexural (dual-cantilever) deformations.

Table 1.1 Specified Ranges of Some Commercial Nonresonance Instruments[a]

Manufacturer	Model	Deformation Modes	Dynamic Force Amplitude	Displacement Amplitude	Frequency Temperature (°C)	Heating Rate (°C/min)
E. I. du Pont deNemours and Company; Wilmington, DL, USA	983 Dynamic mechanical analyzer (DMA)	Flexure, shear		0.1–2 mm Peak to peak	0.001–10 Hz (−150–500)	0.01–50
I Mass, Inc.; Hingham, MA, USA	Dynastat viscoelastic analyzer	Tension-compression, flexure, shear	0.1–100 N	±0.5 μm to ±5 mm	0.01–200 Hz (−150–250)	Various temperature programs
Metravib Instruments; Dardilly, France	Viscoelasticimeter, viscoanalyzer	Tension-compression shear, flexure	0.1–150 N	±0.1 μm to ±0.5 mm	5–1000 Hz (−100–350)	
	Micromecanalyzer	Torsion	Maximum torque 1.5×10^{-2} Nm		10^{-5}–1 Hz (−170–400)	

Manufacturer	Instrument	Deformation mode	Force range	Displacement range	Frequency range (temperature range, °C)	
Polymer Laboratories Ltd., Instrument Division; Loughborough, UK	Dynamic mechanical thermal analyzer (DMTA)	Flexure, shear, tension	Up to 4 N	0.01–0.25 mm	0.01–200 Hz (−150–800)	0.1–20
Rheometrics Inc.; Piscataway, NJ, USA	Solids analyzer RSA II	Tension-compression, flexure, shear	0.01–10 N	$\pm 1\,\mu$m to ± 0.5 mm	0.01–100 rad/s (−150–600)	0.1–50
	Mechanical spectrometer RMS 800	Torsion	Torque 10^{-6}–1 Nm	Up to 0.5 rad angular displacement	10^{-3}–100 rad/s (−150–350)	
Toyo Baldwin Company Ltd.; Tokyo, Japan	Rheovibron viscoelastometer DDV II	Mainly tension	0.003–1 N	0.016, 0.05, and 0.16 mm	0.01–110 Hz (−150–250)	0.5–3
	Rheovibron viscoelastometer DDV III	Mainly tension	0.3–50 N	0.025, 0.08, and 0.25 mm	0.01–110 Hz (−150–300)	0.5–3

[a]This list of available instruments is not exhaustive and does not indicate preference for particular models. The quoted ranges include optional extras and can vary with instrument developments.

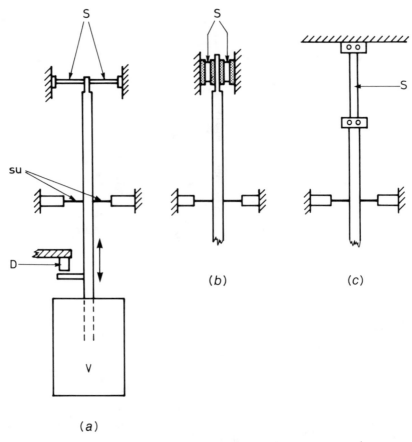

Figure 1.5 Nonresonance method of the type developed by Polymer Laboratories Ltd. The apparatus contains no force transducer and includes a flexible diaphragm drive suspension su. Loading stages are shown for (*a*) flexure, (*b*) shear, and (*c*) tensile deformations.

viscoelastometer, the Dynastat viscoelastic analyzer, the Metravib viscoelastici-meter and viscoanalyzer, the Rheometrics solids analyzer, and equipment developed at the National Physical Laboratory (NPL) [7]. With the second instrument type (Figure 1.5), the applied harmonic forces are determined from the current to the vibrator with the aid of a calibration. Account is then taken of the stiffness of a flexible drive suspension su, which deforms in parallel with the sample. This method is exemplified by the Polymer Laboratories dynamic mechanical thermal analyzer (DMTA) [29] and the Dupont dynamic mechan-ical analyzer (DMA) operating in the nonresonance mode. Although methods for determining phase angles are not often revealed in detail, they can involve measurements of time intervals between the force and displacement signals [7] or measurements of areas within hysteresis loops of force-displacement curves [30].

After allowing for the effects of compliance, inertia, and loss within the apparatus (Section 2.3), the measured dynamic force and displacement yield the components of a complex sample stiffness $k*$ given by

$$k* = F*/u* = k' + ik'' \tag{21}$$

Here the respective force and displacement *experienced by the sample* are given by the real parts of $F* = F_0 \exp[i(\omega t + \delta)]$ and $u* = u_0 \exp(i\omega t)$, F_0 and u_0 being the amplitudes of the force and displacement. It follows that

$$k' = \frac{F_0}{u_0} \cos \delta$$

$$k'' = k' \tan \delta \tag{22}$$

For each mode of deformation the components of an appropriate complex modulus can now be evaluated from the measured k', $\tan \delta$, and sample dimensions. For example, a strip of material can be clamped at each end and subjected to a harmonic uniaxial extension–compression (Figures 1.4a and 1.5c) with a superimposed static extension for samples insufficiently rigid to prevent buckling under compression. It follows from (4), (6), and (22) that

$$E' = k'l/A$$

$$E'' = E' \tan \delta_E \tag{23}$$

where l is the effective sample length between the clamps (including any prestrain) and A is the sample cross-sectional area.

Some nonresonance instruments include loading stages that produce simple shear of two sandwiched blocks of material each bonded to metal endpieces (Figures 1.4b and 1.5b). The dynamic shear modulus components are obtained from

$$G' = \frac{k'h}{2A} \left[1 + \frac{h^2}{l^2} \frac{G'}{E'} \right] \tag{24}$$

and

$$G'' = G' \tan \delta_G$$

where h is the thickness of each block, A is the area of each bonded face, and l is the sample length in the direction of the applied displacement. The second term in the brackets in (24) represents a small correction G' arising from a contribution from bending to the displacement.

With flexural deformations a material strip or beam is often clamped at its center and at each end in the so-called *dual-cantilever mode* (Figures 1.4c and 1.5a). The components of E^* are then evaluated from the relations

$$E' = \frac{k'l^3}{2bh^3} \left[1 + \frac{h^2}{l^2} \frac{E'}{G'} \right] \qquad (25)$$

and

$$E'' = E' \tan \delta_E$$

where b and h are, respectively, the width and thickness of the beam; l is the effective length between the center and each outer clamp; and the second term in the brackets corresponds to a correction for shear deformations that is significant for short, thick beams.

Other modes of harmonic deformation, such as torsion and hydrostatic compression, are employed in studies of solids with the nonresonance method. With torsional equipment of the type produced by Rheometrics (Table 1.1) a harmonic angular displacement is applied and monitored at one end of a solid rod of rectangular or circular cross section, and the torque is measured at the other end by a torque transducer. Other torsional instruments, including the Metravib micromecanalyzer (Table 1.1), apply a torque electromagnetically at the top end of a rod suspended from a torsion wire [31]. The angular displacement is measured electrooptically, and the torque is determined by a calibration from the current supplied to the electromagnet. The torsional nonresonance instruments yield the components of a complex torsional sample stiffness $\Gamma^* = T^*/\theta^*$. Here T^* and θ^* are, in complex form, the torque (moment) and the angular displacement *experienced by the sample*. These are obtained from the measured torque and displacement after allowing for apparatus compliance, inertia, and loss (see Section 2.3). The G^* components are then obtained from the components of the sample stiffness using $\Gamma^* = KG^*/l$, where K is a constant whose value depends on the sample cross-sectional shape and dimensions. For samples of rectangular cross section, with $h/b < \frac{1}{2}$, K equals $bh^3[1 - 0.63(h/b)]/3$ to a good approximation [7].

The components of the bulk modulus K^* for solid polymers were determined from direct measurements of the volume response to applied alternating hydrostatic pressures [32]. For glassy polymers, the K^* components can also be obtained with little error magnification from simultaneous nonresonance measurements of the components of E^* and the complex Poisson's ratio v^* (Section 1.1). This method involves observations of the relative amplitudes and phases of the lateral and longitudinal strains during tensile tests on solid strips [16, 33].

2.2   Range and Accuracy of the Technique

The nonresonance method can be used to study a wide range of materials with storage moduli between about 0.1 MPa and 200 GPa. The selection of an appropriate instrument and mode of deformation depends partly on the particular modulus required, hence, on the data application, and on conveniently measured magnitudes of force and deformation. Compliant materials such as rubber having G' or E' in the range from 0.1 to 10 MPa are best studied either in tension–compression or in simple shear. Dynamic tensile–compressive deformations of long, rubberlike strips necessitate the application of prestrains to avoid buckling and yield or require small force amplitudes F_0. These disadvantages can be minimized by testing short, thick samples under compression between parallel end plates. However, because of the influence of end constraints in restricting lateral sample displacements, (23) is no longer accurately valid, and results are strongly dependent on sample geometry [30]. Shear tests also avoid the requirement for prestrains and, since $2A/h$ values in shear are considerably larger than A/l values used in tension, they yield relatively high F_0 values for a given displacement or strain amplitude.

Simple shear deformations are less suitable for stiffer materials since the high force levels associated with measurable deformations demand strong bonding and a very robust machine. Thus materials with G' or E' in the range from 10 MPa to 10 GPa are more conveniently studied in tension, flexure, or torsion; bear in mind that clamping errors should be significant and can be corrected for by methods considered in Section 2.4. These materials include partially crystalline polymers and amorphous polymers in the glassy state or at temperatures within the glass–rubber transition region. Nonresonance tensile tests are particularly well suited to the characterization of such materials in the form of films or monofilaments.

Nonresonance measurements in flexure or torsion can be used to determine E' or G' of very rigid materials with moduli above 10 GPa, including inorganic glasses, metals, ceramics, and composites. For fiber-composite materials, measured E' values from flexural tests may be highly anisotropic, as discussed in Section 1.2, and the longitudinal E_1' values are subject to relatively large corrections for shear deformation [see (25)] because of the high E_1'/G_{12}' ratios [7]. Note also that E' values obtained from flexural tests on very rigid materials require appreciable corrections for clamping errors (see Section 2.4). Accurate moduli for these materials are obtained readily from free–free resonance tests (Section 4) and ultrasonic measurements (Section 5), which avoid effects caused by clamping.

The resolution in phase-angle measurement with the nonresonance technique is typically about $0.1°$ ($\tan \delta = 0.002$); but, allowing for spurious sources of error, we estimate that sample $\tan \delta$ values around 0.02 are the lowest that can be determined with reasonable accuracy. The percent error decreases with increasing $\tan \delta$ so that the method is ideally suited to the study of high-loss materials, including amorphous polymers in the glass–rubber transition region ($\tan \delta = 0.1-2$) and uncross-linked polymers in the terminal region of viscoelastic behavior (Section 1.4). To study such materials the method has distinct advantages over the torsion pendulum, resonance, and ultrasonic methods, but it is less suited than the resonance method for studying rigid, low-loss materials with $\tan \delta$ below about 0.01.

Other advantages of the nonresonance method concern its ability to determine the dependence of dynamic properties on frequency and strain amplitude over relatively wide and continuous ranges. Most commercial instruments operate at several fixed frequencies that typically range from 0.01 to 100 Hz. Below 0.01 Hz electronic recording problems arise that are associated with signal drift. The upper frequency limit is governed by resonances in the apparatus and sample as discussed in Section 2.3. Reliable data up to 1000 Hz can be obtained through appropriate design of equipment and by correcting for or avoiding resonance effects. Flexible suspensions such as the torsion wires used in some torsional instruments [31] can produce low resonance frequencies, which limit measurements to frequencies below 10 or 1 Hz.

The dynamic strain range depends on factors such as the vibrator displacement, transducer sensitivity, and mode of deformation. Displacement amplitudes between $10\,\mu m$ and $5\,mm$ can be readily applied and accurately measured. In tensile tests with typical sample dimensions, these displacements yield strains between about 10^{-4} and 10^{-1}, while in simple shear they produce shear strains that range from 10^{-2} to 1. With flexural and torsional deformations the strains are nonuniform, attaining maximum values along the sample faces [7], and they are often small relative to those in simple tension or shear. For example, with the clamped flexural vibrations discussed previously [see (25)] root-mean-square strain amplitudes can be evaluated from $\langle\epsilon_0^2\rangle^{1/2} = \epsilon_{0,max}/3 = hu_0/l^2$, where $\epsilon_{0,max}$ is the maximum strain amplitude. Taking $l = 50\,mm$, $h = 1\,mm$, and $u_0 = 10\,\mu m$ we obtain $\langle\epsilon_0^2\rangle^{1/2} = 4 \times 10^{-6}$. For materials that show a dependence of moduli on dynamic strain, a sinusoidal applied strain will not generally produce a sinusoidal force and vice versa. In this situation (23)–(25) are not strictly valid for the data analysis, but they can be employed to obtain apparent moduli from amplitude and phase measurements on the distorted stress or strain signals.

2.3   Effects of Apparatus Compliance, Inertia, and Loss

In calculating components of the complex sample stiffness k^* using (21) and (22) u_0, F_0, and δ for the sample frequently differ from the respective measured quantities u_{0m}, F_{0m}, and δ_m, which are obtained from the displacement and force transducer (or drive current) signals. These differences arise from contributions to the measured displacements and forces generated in different parts of the apparatus. With equipment of the type illustrated in Figure 1.4, an allowance for instrument compliance can be made by considering the proposed model in Figure 1.6a [7]. By neglecting the presence of mass in the equipment, and hence inertial forces, this model is applicable to the apparatus response at low frequencies. In the diagram k_1 represents the stiffness of the force transducer, which is often considered to be the main source of apparatus compliance. The symbols k_2 and k_3 denote the stiffnesses of connections at each end of the sample; and, since the rods connecting the sample to the force transducer and vibrator table are usually rigid, they may largely reflect displacements of the clamping devices or associated bolted joints. Assuming that losses in the equipment are negligible we take k_1, k_2, and k_3 to be real. Since inertial forces are

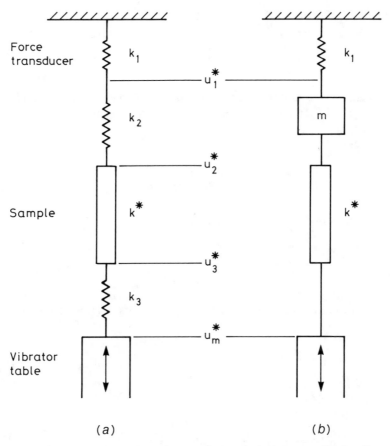

Figure 1.6 Models applicable to the equipment of Figure 1.4 that account for the effects of (a) machine compliance and (b) the inertia and compliance associated with the force transducer resonance.

also neglected, then, using complex notation, the measured force F_m^* equals the force F^* on the sample and other components and is given by

$$k_1 u_1^* = k_2(u_2^* - u_1^*) = k^*(u_3^* - u_2^*) = k_3(u_m^* - u_3^*) \tag{26}$$

where the displacements u_1^*, u_2^*, u_3^*, and u_m^* refer to the positions indicated in Figure 1.6a. The measured stiffness k_m^* then becomes

$$k_m^* = F^*/u_m^* = k_1 u_1^*/u_m^* \tag{27}$$

Using the identity

$$u_m^* = u_1^* + (u_2^* - u_1^*) + (u_3^* - u_2^*) + (u_m^* - u_3^*)$$

together with (26) and (27) it follows that

$$\frac{1}{k_m^*} = \frac{1}{k_1} + \frac{1}{k_2} + \frac{1}{k_3} + \frac{1}{k^*} \tag{28}$$

If the sample is now replaced by an elastic component that is much stiffer than k_1, k_2, and k_3, then k^* can be taken as real and infinite, and the measured stiffness k_∞ becomes

$$\frac{1}{k_\infty} = \frac{1}{k_1} + \frac{1}{k_2} + \frac{1}{k_3} \tag{29}$$

which, when substituted into (28) yields

$$k^* = \frac{k_m^*}{1 - (k_m^*/k_\infty)} \tag{30}$$

After separating the real and imaginary components in (30), we finally obtain

$$k' = \frac{|k_m|[\cos \delta_m - (|k_m|/k_\infty)]}{1 - 2(|k_m|/k_\infty) \cos \delta_m + (|k_m|/k_\infty)^2} \tag{31}$$

and

$$\tan \delta = \frac{\tan \delta_m}{1 - (|k_m|/k_\infty \cos \delta_m)} \tag{32}$$

where $|k_m| = F_{0m}/u_{0m}$.

In applying equations (31) and (32) to the correction of dynamic data, k_∞ can be determined from measurements on very rigid metal samples in low-frequency tensile or shear tests. Except for high-loss materials, the corrected dynamic stiffness will be higher than the measured one, the magnitude of the correction increasing with $|k_m|/k_\infty$ and hence with sample stiffness. Corrected $\tan \delta$ values are also higher than the measured $\tan \delta_m$ particularly for samples of high loss and/or high stiffness [34].

The influence of inertial forces on the force transducer signal (Figure 1.4) is generally complicated. However, if k_2 and k_3 are each assumed to be much larger than k_1 (or $k_\infty \approx k_1$), then these forces may be allowed for on the basis of the model in Figure 1.6b, where m is the mass of the assembly between the transducer and sample. With this model the force on the sample differs from the measured force $k_1 u_1^*$ and is given by

$$k_1 u_1^* + m d^2 u_1^*/dt^2 = k^*(u_m^* - u_1^*) \tag{33}$$

where u_1^* and u_m^* are indicated on Figure 1.6b. Since $d^2u_1^*/dt^2 = -\omega^2 u_1^*$ we then obtain

$$k^* = \frac{[1 - (m\omega^2/k_1)]k_m^*}{1 - (k_m^*/k_1)} \tag{34}$$

where $k_m^* = k_1 u_1^*/u_m^*$. From (34) expressions for k' and $\tan\delta$ are obtained that are identical to those derived for the Rheovibron [35] and the Dynastat [36] instruments, and measured factors $1 - (m\omega^2/k_1)$ were employed in the correction of k' values for system inertia [36]. According to this factor errors of less than 1% in k' are obtained if $m\omega^2/k_1 \leqslant 0.01$, a condition satisfied if the frequency f (measured in hertz) is less than $0.1f_r$, where the transducer resonance frequency f_r equals $(1/2\pi)(k_1/m)^{1/2}$. On this basis the maximum frequencies below which inertial effects can be avoided are typically between about 20 and 150 Hz. Note that when $k_1 = k_\infty$ and $m\omega^2/k_1 \ll 1$, the expressions for k' and $\tan\delta$ reduce to (31) and (32), respectively. Measurements on several instruments [7, 35, 36] suggest that the ratio k_∞/k_1 often lies between about 0.5 and 0.9, so that the assumptions underlying the derivation of (34) may not be accurately valid. At frequencies above 100 Hz other sources of equipment resonance in addition to the transducer resonance can also become significant. Correction procedures were proposed for these effects and for the influence of standing wave resonances in flexible samples [7]. We conclude that inertial effects are best avoided by working at sufficiently low frequencies, employing (31) and (32) to correct for the apparatus compliance.

The Polymer Laboratories DMTA instrument (Figure 1.5), and other equipment of this type, can be analyzed in terms of the model in Figure 1.7. In this diagram m is the mass of the drive shaft, which dominates the inertial forces. The apparatus compliance and loss are each accounted for by assigning a complex stiffness k_{su}^* to the flexible suspension. Other parts of the system are designed to be sufficiently rigid compared with the sample so as to exhibit negligible deformation under the maximum available drive force. The displacements experienced by the sample and suspension are then both equal to the measured displacement u_m^*; the measured drive force is given by

$$F_m^* = md^2u_m^*/dt^2 + k^*u_m^* + k_{su}^*u_m^* \tag{35}$$

and the complex measured stiffness by

$$k_m^* = F_m^*/u_m^* = -\omega^2 m + k^* + k_{su}^* \tag{36}$$

Separating the real and imaginary components in (36) it follows that

$$k' = |k_m| \cos\delta_m - |k_{su}| \cos\delta_{su} + \omega^2 m \tag{37}$$

$$k'' = |k_m| \sin\delta_m - |k_{su}| \sin\delta_{su} \tag{38}$$

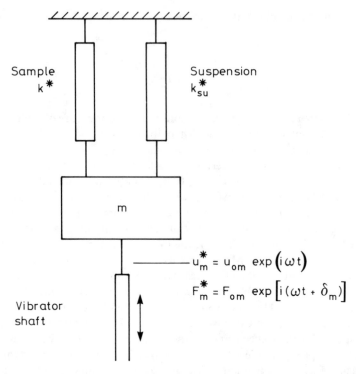

Figure 1.7   Model of the apparatus in Figure 1.5 that accounts for the stiffness and loss of the suspension and the inertia of the drive mass.

and

$$\tan \delta = \frac{|k_m| \sin \delta_m - |k_{su}| \sin \delta_{su}}{|k_m| \cos \delta_m - |k_{su}| \cos \delta_{su} + \omega^2 m} \tag{39}$$

where $|k_{su}|$ and δ_{su} are the respective magnitude (F_{0su}/u_{0su}) and phase angle associated with k_{su}^*.

Values of k', k'', and $\tan \delta$ for the sample are thus obtained using (37)–(39) from the measured $|k_m|$ and δ_m if $|k_{su}|$, δ_{su}, and m are known. Assuming that the suspension stiffness and damping are independent of frequency, we derive both $|k_{su}| \cos \delta_{su}$ and m using (37) from measurements of $|k_m| \cos \delta_m$ in the absence of a sample ($k' = 0$) at two known frequencies. At least one of these frequencies must be sufficiently high such that $\omega^2 m$ is not negligible compared with $|k_{su}| \cos \delta_{su}$. The value of $|k_{su}| \sin \delta_{su}$ is similarly determined using (38) from a measurement of $|k_m| \sin \delta_m$ at any frequency, again with no sample present ($k'' = 0$).

The above methods can also be employed to analyze torsional nonresonance data if the stiffnesses $|k_m|$, k_∞, and $|k_{su}|$ are replaced by analogous torsional rigidities $|\Gamma_m|$, Γ_∞, and $|\Gamma_{su}|$, respectively; and an appropriate moment of inertia I

is substituted for the mass m. Equations analogous to (31), (32), and (34) could then be used to account for machine compliance and inertia for those methods that employ a torque transducer to measure the restoring torques in samples subjected to harmonic angular displacements. Similarly, equations analogous to (37) and (39) will allow for the suspension stiffness and loss and the system inertia for those techniques employing a torsion-wire suspension.

2.4   Corrections for Clamping Errors in Tension and Flexure

The determination of E' from dynamic tensile or flexural tests is subject to errors originating from some deformation of material contained within the grips. Because of this effect the quantity l in (23) or (25) is regarded as an effective length that slightly exceeds the measured grip separation l_m by an amount Δl. By investigating the dependence of k' or the apparent modulus on l_m, an estimate of Δl and a corrected E' value can be obtained.

2.4.1   Tensile Deformations

Assume that the length correction Δl is a constant independent of l_m over the range of l_m investigated; then equation (23) for the calculation of E' from tensile tests becomes

$$\frac{1}{k'} = \frac{l}{E'A} = \frac{l_m + \Delta l}{E'A} \tag{40}$$

If values for k' are obtained as l_m for a specimen is systematically varied, then a plot of $1/k'$ versus l_m should be linear, allowing Δl to be obtained from the intercept at $1/k' = 0$ and yielding E' from the slope $(1/E'A)$. Alternatively, a plot of l_m/E'_a $(= A/k')$ versus l_m, where E'_a is the apparent modulus uncorrected for clamping effects, should be linear with intercept Δl at $l_m/E'_a = 0$ and slope $1/E'$. Figure 1.8 shows a plot of $1/k'$ against l_m for a polymer blend (see Section 2.5) containing 50 wt% of both polycarbonate (PC) and linear, low-density poly(ethylene) (LLDPE). For maximum accuracy in the extrapolation it is important that the temperature and strain amplitude are held constant and that corrections are made for apparatus compliance at each l_m. The respective width and thickness of the blend sample was 10 and 3 mm, and the data were obtained on the NPL equipment [7] at 23°C, a frequency of 1 Hz, and constant strain amplitude of 1.3×10^{-4}. Corrections to k' between 4 and 10% for apparatus compliance were first applied using the measured k_∞ value of 1.0×10^7 N/m. The plot in Figure 1.8 is accurately linear and yields a Δl value of 6 mm and storage modulus E' of 1.34 GPa. Length corrections in tension usually lie between about 2 and 12 mm and tend to increase with increasing E' and hence with decreasing temperature, but the exact dependence on sample modulus and geometry is not established. Furthermore, because of the rather long extrapolations, Δl values are reproducible only to about ± 2 mm, which implies that an accuracy of about $\pm 2\%$ is obtained by this method.

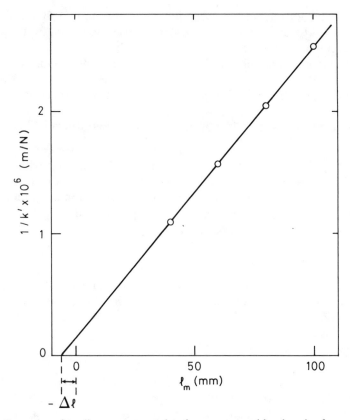

Figure 1.8 Correction of tensile nonresonance data for errors caused by clamping for a polymer blend sample containing 50 wt% of PC and LLDPE at 23°C and 1-Hz frequency.

2.4.2 Flexural Deformations

In deriving length corrections applicable to results from dynamic flexural tests, we first write (25) in the form

$$
\frac{2bh^3}{k'\left[1 + \dfrac{E'}{G'}\left(\dfrac{h}{l_m + \Delta l}\right)^2\right]} = \frac{(l_m + \Delta l)^3}{E'} \tag{41}
$$

If Δl on the left-hand side of (41) is now ignored, assuming that the shear correction term is small compared with unity, then the equation becomes

$$
\frac{l_m}{E_a'^{1/3}} = \frac{l_m + \Delta l}{E'^{1/3}} \tag{42}
$$

where E'_a is the apparent modulus evaluated from (25) assuming that $l = l_m$. According to (42) a plot of $l_m/E'^{1/3}_a$ versus l_m should be linear, having an intercept Δl at $l_m/E'^{1/3}_a = 0$ and a slope $1/E'^{1/3}$. Figure 1.9 shows such a plot for a sample of PC (Section 2.5 below) with $b = 6$ mm and $h = 4$ mm studied in the dual-cantilever mode with the Polymer Laboratories DMTA instrument. The measurements were made at 23°C, at a constant frequency of 1 Hz and constant displacement amplitude of 15.6 μm. The average strain amplitude thus varied with l_m from 2.4×10^{-4} to 9.8×10^{-4}, but it was considered sufficiently small to maintain linear behavior. Clamping pressures were also optimized by employing a torque screwdriver to maximize the recorded E'_a values. From the linear plot in Figure 1.9 we obtain $\Delta l = 1.1$ mm and $E' = 2.46$ GPa. With the Polymer Laboratories flexural method, values of Δl frequently range from 0.7 to 1.5 mm and increase with factors such as increasing thickness and modulus and decreasing temperature, which together increase the sample stiffness. For usual sample lengths between 7 and 16 mm it follows that corrections to E' for clamping are often between about 13 and 65%. These relatively large corrections clearly relate to the short specimen lengths required for accurate temperature control and analysis of small quantities of material. The resulting E' values also tend to be somewhat lower than the corrected E' values from dynamic tensile tests on long samples of the same material. Assuming greater reliability of E' values obtained on larger specimens, these values may serve as a basis for further corrections or for setting instrument calibrations to allow for clamping errors.

2.5   Temperature-Dependent Studies of a Polymer Blend

Several commercial, automated nonresonance instruments yield rapid numerical and graphical records of dynamic properties over wide temperature ranges at several frequencies. As an illustration of the temperature dependence of dynamic properties, some data are shown here that were obtained at the NPL as part of a cooperative program on polymer blends associated with the Versailles Agreement on Advanced Materials and Standards (VAMAS). The materials investigated were LLDPE; PC; and blends of these two polymers containing 25, 50, and 75 wt% PC, respectively. Extruded sheets of these materials were supplied by the Industrial Materials Research Institute of the National Research Council of Canada. The dynamic tensile and flexural data reported here were obtained on strips cut parallel to the extrusion direction from the slightly anisotropic sheets.

Values of E' were obtained from tensile measurements at -100, 23, and 100°C and at 1 Hz on samples of each material having $b = 10$ mm and $h = 3$ mm after applying length corrections as exemplified previously (Figure 1.8). For the flexural measurements the Polymer Laboratories equipment was employed at 1 Hz in the dual-cantilever mode on samples with l, b, and h equal to 14, 6, and 4 mm, respectively. As in Figure 1.9, length corrections (between 20 and 30%) were obtained for each material at 23°C. These corrections were assumed to be independent of temperature in the subsequent correction of E' data obtained by

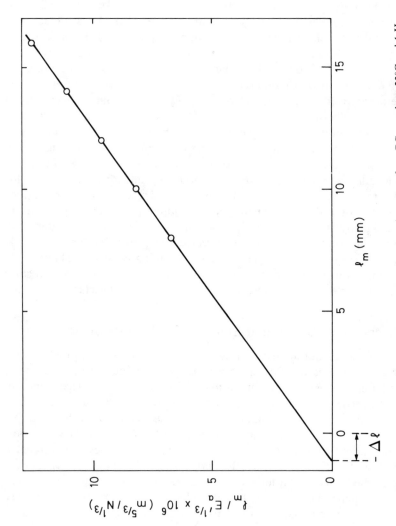

Figure 1.9 Correction of flexural nonresonance data for clamping errors for a PC sample at 23°C and 1-Hz frequency.

scanning at $2°C/min$ between -130 and $180°C$. Additional adjustments around 10% were then required, based on factors averaged over the temperature range, to achieve close agreement with the tensile data. The resulting E' values shown in Figure 1.10a agree within 5% over the whole temperature range with tensile results from the Rheovibron DDV-III viscoelastometer that were obtained independently in two other laboratories [37, 38].

Although the E' versus temperature plots clearly reveal the major transitions, including the T_g of PC around $145°C$ and the onset of melting of LLDPE in the region of $125°C$, the secondary relaxation regions are more evident in the variation of $\tan \delta$ with temperature (Figure 1.10b). For example, in addition to the melting temperature T_m of LLDPE, the $\tan \delta$ temperature plot shows peaks labeled α, β, and γ in the region of 70, -25, and $-110°C$, respectively. The α peak is thought to result from motions of restrained molecular segments close to the crystal surfaces; the β peak is assigned to the micro-Brownian segmental motions associated with T_g of LLDPE; and the γ peak is assigned to localized motions of short $—CH_2—$ sequences [5]. For PC the α and β peaks at 149 and $-90°C$, respectively, correspond to long-range motions related to T_g and a more localized secondary process [5]. It should be emphasized that the magnitudes, shapes, and locations of these peaks are close to those observed with the Rheovibron instrument [37, 38]. For the blend with 50 wt % PC and LLDPE the locations of the different peaks are also very close to those found for the individual blend constituents. This result illustrates the two-phase structure of the immiscible blend and it is consistent with morphological evidence. Systems that are miscible on a scale of 10 nm or less exhibit a single, somewhat broadened T_g peak caused by the merging of the peaks associated with the separate components [24].

3 TORSION PENDULUM

In Section 2.2 it was mentioned that torsional deformations are more suitable than simple shear for determining the dynamic shear properties of rigid materials as they avoid the high force levels associated with measurable deformations. The torsion pendulum is particularly convenient for such determinations, yielding G' and $\tan \delta_G$ from the natural frequency and decay rate of free torsional oscillations of a sample with added inertia. Although less versatile than nonresonance techniques, this method allows more accurate measurements of loss factor to be made on low-loss materials ranging from soft solids [39] to rigid metals [40] within a restricted frequency range (0.1–10 Hz). Highly automated instruments are now commercially available (Table 1.2) that enable efficient determinations of dynamic shear properties over wide temperature ranges and some of which conform to existing standards [41].

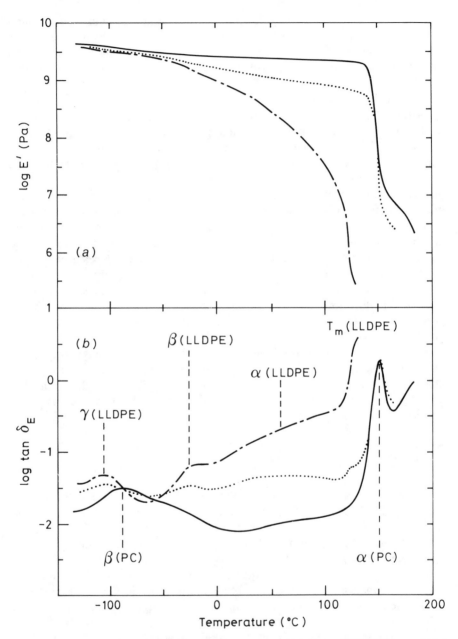

Figure 1.10 Variation of (a) E' and (b) tan δ_E with temperature for PC (———), LLDPE (———), and a blend containing 50 wt% PC and LLDPE (···). Results were obtained with the Polymer Laboratories dual-cantilever flexural apparatus at 1-Hz frequency, an average strain amplitude 3.2×10^{-4}, and a heating rate of $2°C/min$. Applied corrections are discussed in the test.

Table 1.2 Specified Ranges of Some Commercial Torsion Pendulum Instruments[a]

Manufacturer	Model	Moment of Inertia (kg mm²)	Frequency (Hz)	Temperature (°C)	Heating Rate (°C/min)
Brabender OHG; Duisburg, Federal Republic of Germany	Torsionautomat torsion pendulum	40–200	0.01–10	−180–400	0.05–2
Myrenne GmbH; Roetgen, Federal Republic of Germany	Automatic torsion pendulum ATM 3	5.5–2373	0.1–10	−180–300	Selectable heating program
Plastics Analysis Instruments, Inc;; Princeton, NJ	Torsion braid analyzer (TBA)		0.05–10	−190–400	0.05–5
Zwick GmbH; Ulm, Federal Republic of Germany	Torsiomatic Zwick 5204 torsion pendulum		0.5–25	−180–400	1–3

[a]This list of available instruments is not exhaustive and does not indicate preference for particular models.

3.1   Apparatus and Theory

The basic design for a torsion pendulum [42], illustrated schematically in Figure 1.11a, was employed in the construction of recent commercial instruments, including the Zwick Torsiomatic torsion pendulum and the torsion braid analyzer (TBA) [43]. The sample S is conveniently in the form of a strip of rectangular cross section or a cylindrical rod. However, with the TBA, which provides only relative material properties, the sample comprises a supporting glass braid impregnated with a small quantity of the material to be analyzed. The sample is clamped at its upper end to a rigid support, and an inertia member I, which can be a disk or a horizontally supported rod, is freely suspended from the lower end of the sample. Torsional oscillations are induced by the

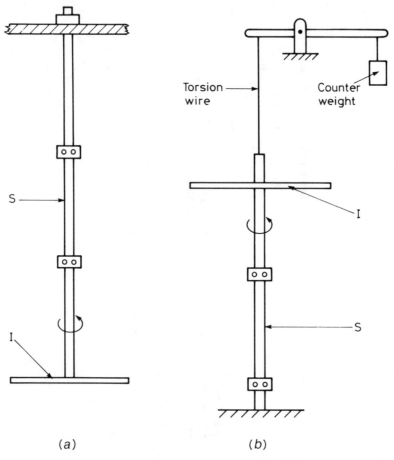

(a)   (b)

Figure 1.11   Diagrams of the torsion pendulum illustrating (a) the basic design and (b) a design incorporating a torsion-wire suspension and method for counterbalancing the mass of the inertia member I.

application of a torque pulse through either the upper clamp or the inertia member; and the time dependence of the angular displacements is monitored by optical, electrical, electrooptical, or magnetic devices [42–44].

If we ignore the influence of the tensile load on the restoring torque, and extraneous effects such as air damping, we can derive the equation of motion for the system by balancing the inertial forces with the viscoelastic forces expressed in terms of Boltzmann's superposition principle [45]. On this basis it can be shown that free damped harmonic motions are governed by

$$I\left(\frac{d^2\theta}{dt^2}\right) + \left[\frac{KG^*(\omega^*)}{l}\right]\theta = 0 \tag{43}$$

where θ is the angular displacement, I is the inertia of the system, K is the shape constant discussed in Section 2.1, and $G^*(\omega^*)$ is a complex shear modulus for the *complex* frequency $\omega^* = \omega + i\alpha$. The solution to (43) is then

$$\theta = \theta_0 \exp(i\omega^* t)$$

$$= \theta_0 \exp\left(-\frac{\omega\Lambda t}{2\pi}\right)\exp(i\omega t) \tag{44}$$

which represents a damped sinusoid as shown in Figure 1.12 with angular frequency ω, amplitude θ_0 at $t = 0$ and decay rate specified by the logarithmic decrement $\Lambda = \ln(\theta_n/\theta_{n+1}) = 2\pi\alpha/\omega$. For low damping $(\tan\delta_G \leqslant 0.1)$, $G^*(\omega^*) \approx G^*(\omega)$ and substituting (44) into (43) we obtain

$$G' = \frac{4\pi^2 I l f^2}{K}$$

$$G'' = \frac{4\pi I l \Lambda f^2}{K} \tag{45}$$

and

$$\tan\delta_G = \Lambda/\pi$$

where the frequency of oscillation f (measured in hertz) $= \omega/2\pi$. The storage modulus G' can thus be evaluated from measurements of f if I and the sample dimensions are known; $\tan\delta_G$ is obtained directly from the measured Λ.

With the simple torsion pendulum described earlier (Figure 1.11a) various methods were devised for suppressing the lateral movements that may accompany the torsional oscillations. These include a magnetic stabilizer [43], a needlepoint bearing [46], and a gas-lubricated bearing [44]. The tensile stress on the specimen resulting from the suspended inertia member can also yield errors in G' by virtue of its influence on the restoring torque [5]. These errors are

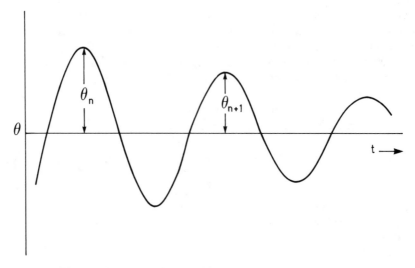

Figure 1.12 Notation for the free decay of torsional oscillations.

generally significant only for flexible materials ($G' \leqslant 10\,\text{MPa}$) where orientation and tensile creep produce additional problems. The effects of both lateral movement and tensile stress can be eliminated if the inertia member is suspended from an air bearing and is located above the sample. Equipment developed at the NPL [7] employs this technique together with a device for counterbalancing the mass of the lower clamp and a linear bearing attached to the clamp that allows for longitudinal thermal expansion and contraction of the specimen.

Tensile stresses can also be eliminated or controlled if the inertia member is suspended by a torsion wire above the sample and a mechanism included for counterbalancing its mass. This well-known method [42, 47], illustrated in Figure 1.11b, is used in commercial instruments such as the Brabender Torsionautomat apparatus. The Myrenne automatic system ATM 3 similarly excludes tensile stresses by suspending the inertia member between stretched springs below the specimen. These methods also reduce lateral movements of the pendulum and the added stiffness of a torsion wire allows measurements to be made on flexible high-loss materials by maintaining a low overall level of damping. The effect of the wire is accounted for by adding a contribution $\Gamma^*_{su}\theta$ to the restoring torque on the left-hand side of (43), where Γ^*_{su} is the complex torsional rigidity of the wire. Ignoring air damping and friction, we obtain from the equation

$$G' = 4\pi^2 Il(f^2 - f_0^2)/K$$

$$G'' = 4\pi Il(\Lambda f^2 - \Lambda_0 f_0^2)/K \tag{46}$$

$$\tan \delta_G = (\Lambda f^2 - \Lambda_0 f_0^2)/\pi(f^2 - f_0^2)$$

where f and Λ are the frequency and logarithmic decrement measured with the sample and torsion wire both present and f_0 and Λ_0 are the corresponding quantities determined for the wire alone. It is clear from (46) that, as the rigidity of the wire suspension increases, the influence of sample properties is reduced, which results in a decreased accuracy of calculated properties. However, greater accuracy in measurements can be achieved as the motion is sustained longer by the overall decrease in damping.

3.2   Range, Accuracy, and Correction of Errors

With appropriate levels of inertia and machine rigidity, the torsion pendulum can be used to measure shear moduli for many types of solid material ranging from soft solids such as rubbers ($G' \approx 1\,\text{MPa}$) to metals ($G' = 10-100\,\text{GPa}$). Values of G'_{12} and G'_{23} for carbon-fiber composites were also determined [7]. As discussed previously [7], accurate modulus measurements require a precise determination of the inertia I and the availability of uniform samples with accurately known dimensions. Because of the dependence of K on h^3, errors associated with thickness measurements for samples of rectangular cross section are magnified in the G' values.

The effective sample length l in (43) also differs in general from the measured length between the clamps l_m. This difference arises partly from some deformation of material within the clamps where l_m underestimates the effective length. For samples of rectangular cross section, restraints imposed by the clamps on the warping of cross sections [7] raise the torsional rigidity above that given by the use of l_m in (43) so that l_m overestimates the value of l. The correction Δl to l_m is positive or negative depending on the predominant source of error, and it can be determined by writing $l = l_m + \Delta l$ in the expression for G' in (45). It follows that a plot of $1/f^2$ versus l_m is linear if other quantities are constant, and extrapolation to $1/f^2 = 0$ allows us to obtain Δl from the intercept on the l_m axis. Experiments suggest that Δl tends to be negative for samples of rubber, a glassy polymer [poly(methyl methacrylate)], and longitudinally reinforced plastic with rectangular cross sections, requiring negative corrections to G' between about 2 and 8%. Positive corrections were obtained for a transversely reinforced composite.

Methods for minimizing errors in $\tan \delta_G$ caused by noise or drift on the displacement signals were discussed, as were corrections for air damping and the presence of friction in the apparatus [7]. With the NPL apparatus, contributions to $\tan \delta_G$ from air damping and the presence of friction increased from about 10^{-5} for rigid plastics ($G' = 1-10\,\text{GPa}$) to 4×10^{-3} for compliant rubbers. If these sources of error are not eliminated by employing vacuum environments and eliminating bearings, then the lowest $\tan \delta_G$ values that can be determined with good accuracy are around 0.001 providing that with the more compliant materials corrections are applied or a torsion wire or other added stiffness is present to reduce the extraneous damping contribution. A maximum reliable $\tan \delta_G$ level of about 0.1 is determined by the validity of the theory outlined earlier, but values in excess of 1 can be determined if an added stiffness member is included to maintain the net $\tan \delta$ below 0.1. Materials with loss factor above 1 may also be studied with the TBA, but quantitative $\tan \delta_G$ values for the materials cannot be evaluated with this technique.

Torsional deformations produce nonuniform shear strains, which, for samples of rectangular cross section, have a maximum value $\epsilon_{max} = h\theta/l$ along the center of the wider surface. The International Organization for Standardization standard 537 [41] recommends that θ should be no greater than 1.5°, which corresponds to a maximum strain around 3×10^{-4} for $h/l = 0.01$ and should generally ensure linear behavior. Strain variations ranging from 2×10^{-4} to 1×10^{-3} have, however, revealed nonlinear effects because of internal damage in fiber composite materials [48].

The frequency range of 0.1–10 Hz lies sufficiently below the torsional resonance frequencies of samples without added inertia (Section 4.2) and avoids problems such as drift associated with long recording times. At constant I the frequency will, of course, vary with G' and exhibit significant variations with temperature over the range of -180 to $+300°C$, which is often available. Facilities for automatically varying I to maintain constant frequency are provided, for example, with the Myrenne instrument, and aid investigations of the frequency dependence of dynamic shear properties over the limited range available.

4 AUDIOFREQUENCY RESONANCE TECHNIQUES

At frequencies between about 20 Hz and 20 kHz accurate values of storage moduli and loss factors for rigid, low-loss materials can be determined by resonance methods. These techniques involve the application of a harmonic force or moment of constant amplitude at some point along a strip of the material and monitoring of the vibration amplitude as a function of the forcing frequency. At the so-called *resonance frequencies* f_{rn}, maxima are observed in the vibration amplitude (Figure 1.13) corresponding to the appearance of standing waves in the sample. Here the mode number n has values of 1, 2, 3, and so on; the value 1 represents the fundamental natural vibration mode, and higher numbers specify the successive harmonics or overtones. According to theory the storage modulus at frequency f_{rn} can be obtained from measurements of f_{rn} and the sample dimensions; and $\tan \delta$ is approximately equal to $(f_{2n} - f_{1n})/f_{rn}$, where $f_{2n} - f_{1n}$ is the resonance peak width at an amplitude $1/\sqrt{2}$ times the maximum amplitude (Figure 1.13). The loss factor is alternatively determined from the decay rate of the vibration amplitude after removal of the driving force.

The average strain amplitudes in standing-wave resonance tests are typically about 10^{-6} so that problems associated with the nonlinearity of dynamic properties do not arise. Clamping errors can also be avoided by employing free–free vibrational modes of appropriately suspended samples. However, the resonance techniques require samples of sufficiently high rigidity (storage

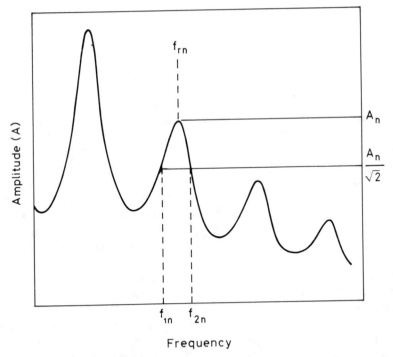

Figure 1.13 Illustration of resonance peaks and definition of a resonance frequency f_{rn} and associated frequencies f_{2n} and f_{1n}.

moduli from 1 to 200 GPa) and low loss (tan δ values below about 0.2) to support their own weight and to exhibit resolved resonance peaks. The audiofrequency properties of flexible, high-loss materials can be determined from resonance tests, however, if the materials are bonded as surface layers to rigid, low-loss substrates [7].

Three types of vibration can be employed with the resonance technique. In order of increasing frequency these are flexural, torsional, and longitudinal. The quantities E' and tan δ_E are obtained from both flexural (20 Hz–10 kHz) and longitudinal (5–50 kHz) resonance measurements. The values of G' and tan δ_G are determined from torsional vibrations at frequencies above 500 Hz. Most resonance instruments, including the commercial Brüel and Kjaer complex modulus apparatus, employ flexural vibrations, although commercial equipment (Dynamod II) from Nene Instruments Ltd, UK, can determine E' and G' for rigid materials from longitudinal, flexural, and torsional resonance frequencies. Details of the flexural technique are presented in the next section, and a brief discussion of torsional and longitudinal resonance is included subsequently. Emphasis is given to an instrument developed at the National Physical Laboratory, where an important aim is to maximize the available frequency range.

4.1   Flexural Resonance

4.1.1   Apparatus

Flexural resonance tests are employed widely to determine the dynamic properties of metals [2, 49, 50], polymers [7, 42, 51], and composite materials [7, 52]. The measurements are made on samples in the form of rectangular strips or cylindrical rods and several methods are used to support the samples and to excite and detect the vibrations. In some techniques the strip is clamped at one end as a cantilever, and clamped-free bending vibrations are induced either by transversely oscillating the clamp [51] or by applying a transverse magnetic alternating force at the free end [7]. The vibration amplitudes are then determined at some point along the sample by electromagnetic, electrostatic, or photoelectric methods. However, for very rigid, low-loss materials such as metals and fiber composites, clamping effects can produce large percentage errors in E' and $\tan \delta_E$, which are not very reproducible and not easily corrected for.

Errors associated with clamping can be eliminated if the sample is suspended within two loops of string and has both ends free. If the strings are located slightly away from the nodes, then free–free vibrational modes can be excited by oscillating one of the loops. The vibrations induced in the second loop can then be detected using various types of pickup [2, 49]. However, significant errors in E' and $\tan \delta_E$ can arise from the added stiffness and energy dissipation caused by the supporting loops. These errors are minimized by locating the strings accurately at the nodes and applying an alternating transverse force at one end of the strip. This technique was employed for studies of polymers [8] and metals [2, 50], and we developed it at NPL for measurements on a range of different materials [7].

The equipment used at NPL is illustrated in Figure 1.14. A rectangular strip of material S with typical dimensions $150 \times 10 \times 2$ mm is suspended horizontally by two short loops of fine nylon filament. The separation between the loops can be varied so that they are located at symmetrically placed nodal points for the different vibrational modes. Transverse (vertical) sinusoidal forces are applied to one end of the sample by an electromagnetic transducer T fed by a variable frequency oscillator Os and power amplifier PA. The resulting alternating magnetic field acts on a small piece of magnetic metal M1 bonded to the specimen. For detecting the vibration amplitude, we employ the probe P of a proximitor, which faces a small conducting mass M2 bonded to the other end of the sample. Operation of the proximitor involves the formation of eddy currents in the probe, which is energized by the proximitor circuit Pr. An output voltage is obtained from Pr that is proportional to the gap separation between the probe tip and the conducting surface for spacings between 0.25 and 1 mm. Transverse bending vibrations in the sample are thus monitored from the cyclic variation of output voltage after amplification by a measuring amplifier MA. A narrow-band tracking filter F, tuned to the drive frequency by the oscillator, is inserted into

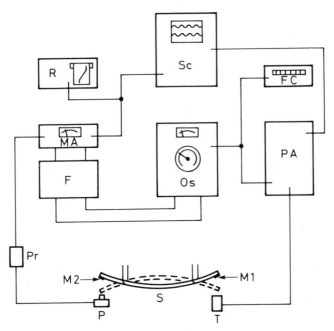

Figure 1.14 Audiofrequency resonance equipment illustrating free–free flexural vibrations of a material strip.

the measuring amplifier. This eliminates extraneous noise and is essential for measurements on the higher modes because of the low signal-to-noise (S/N) ratios. The measuring amplifier is equipped with a meter that records the root-mean-square value of the displacement signal on both a linear and a decibel scale. This is employed when the oscillator is tuned manually to resonance (maximum amplitude) and the resonance peak width is subsequently determined. The relevant frequencies are measured by a frequency counter FC that is connected to the oscillator output. A two-channel oscilloscope Sc can be used to monitor the drive current and displacement signals, and a level recorder R is employed to determine tan δ for very low-loss specimens by recording the free decay of vibration amplitudes after cutting the oscillator output.

The drive transducer was designed with a low inductance to maintain a high level of force to frequencies above 20 kHz. It comprises a 100-turn coil wound around a cylindrical steel core, and it is supplied by the power amplifier with alternating currents up to 5 A. A variable direct current offset up to 5 A serves to polarize the magnet and so yield undistorted harmonic forces with frequencies equal to the oscillator frequency (no frequency doubling). The transducer is liquid cooled to remove the generated heat. To apply corrections for added mass, it is convenient for M1 and M2 to have equal dimensions and mass. For this purpose the two masses were each prepared as thin rectangular pieces from a spring-steel strip. Also note that the sample is placed in an enclosure with facilities for varying both temperature (-30 to $100°C$) and air pressure.

4.1.2 Theory

A detailed discussion was presented of the theory underlying the flexural resonance method [7]. The classical treatment accounts for the tensile-compressive strains normal to the sample cross sections, and it assumes that inertial forces arise only from transverse motions of beam elements. Shear deformations are neglected, as are the rotatory inertial forces caused by the rotational motions of each beam element. For low-loss specimens in the absence of external forces the equation of motion is written

$$\frac{E^*I_a}{\rho A} \frac{\partial^4 w}{\partial x^4} + \frac{\partial^2 w}{\partial t^2} = 0 \tag{47}$$

where w is the transverse displacement at time t of an element of the strip located a distance x from a free end, ρ and A are the respective density and cross-sectional area of the strip, and E^*I_a is its complex flexural rigidity. For rectangular cross sections with width b and thickness h, $I_a = bh^3/12$. Equation (47) yields an infinite number of solutions of the form

$$w_n(x, t) = W_n(x) \exp(i\omega_n^* t) \tag{48}$$

where n equals 1, 2, 3, and so on; and the mode shape $W_n(x)$ and (complex) natural frequency ω_n^* for mode n are related to an eigenvalue α_n, which depends on the boundary conditions at each end of the sample. Values of α_n are listed in Table 1.3 for the first seven modes of a free–free vibrating beam. Figure 1.15 illustrates some mode shapes as a function of x/l, where l is the sample length, calculated from an equation given in [7]. Values of x/l corresponding to the nodal positions $[W_n(x) = 0]$ are also included in Table 1.3. These values are

Table 1.3 Eigenvalues α_n and Nodal Positions (x/l) for a Free–Free Vibrating Beam

n	α_n	x/l				
1	4.730	0.224				
2	7.853	0.132	0.500			
3	10.996	0.094	0.356			
4	14.137	0.074	0.277	0.500		
5	17.279	0.060	0.227	0.409		
6	20.420	0.051	0.192	0.346	0.500	
7	23.562	0.044	0.166	0.300	0.433	

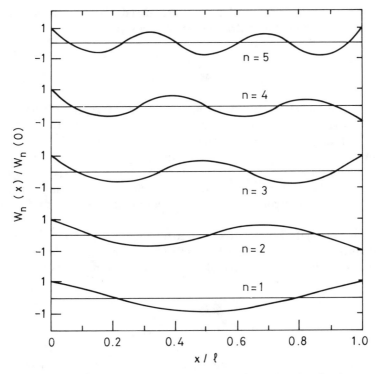

Figure 1.15 Theoretical natural-mode shapes in flexure for a free–free beam.

required for setting the support loops at the nodes during flexural resonance tests.

When a harmonic force is applied laterally to one end of the sample, the transverse displacements can be analyzed by adding an external force term to (47). The solution of the resulting equation [7] shows that for loss factors of less than about 0.1, a displacement amplitude–frequency plot exhibits resolved peaks as in Figure 1.13 at resonance frequencies f_{rn} approximately equal to the real part $\omega'_n/2\pi$ of the natural frequency. This theory also yields a relationship by which E' can be evaluated from the measured f_{rn}. After including correction factors for the effects of added mass $(1 + \psi_m)$ and of shear and rotatory inertia $(1 + \psi_{sr})$, we write the equation

$$E' = \frac{48\pi^2 M l^3 f_{rn}^2}{bh^3 \alpha_n^4} (1 + \psi_m)(1 + \psi_{sr}) \qquad (49)$$

where M is the mass of the sample without added masses and b is the sample width.

The term ψ_m in (49) follows from considerations [7] of the change in kinetic energy caused by the drive and detector masses bonded at opposite ends of the strip (Figure 1.14). The calculations were based on the classical model, neglecting rotatory inertia contributions; and they yielded

$$\psi_m = \frac{4(m_1 + m_2)}{M} \left\langle \frac{W_n^2(x)}{W_n^2(0)} \right\rangle \tag{50}$$

where $(m_1 + m_2)$ is the magnitude of the two added masses and $\langle W_n^2(x)/W_n^2(0) \rangle$ is the mean square value of the displacement ratio $W_n(x)/W_n(0)$; $W_n(0)$ is the displacement amplitude at $x = 0$. The averages were evaluated over the small length Δ (about 5 mm) that each mass extends along the strip from the respective end. Values of these averages are listed in Table 1.4 for various values of Δ/l. Note that the masses employed are usually about 1% of the sample mass and have Δ/l approximately equal to 0.03, which requires less than 4% corrections to E'.

For isotropic materials, a simple but adequate approximation to ψ_{sr} in (49) was obtained by Nederveen and Schwarzl [53] from Timoshenko's equation [54], which includes the effects of both shear and rotatory inertia. This approximation is written

$$\psi_{sr} = \frac{\alpha_n^2 h^2}{12 l^2} \left[1 + \frac{E'}{\kappa_s G'} \right] \tag{51}$$

where κ_s is a constant that accounts for the nonuniformity of shear stresses on each cross section. A value of 0.87 is generally used for κ_s; and E'/G' can be replaced by $2(1 + v)$, where v is Poisson's ratio. It follows from (51) that corrections for shear deformation and rotatory inertia increase with α_n, hence, with mode number n, and also with the thickness-to-length ratio. For a typical l/h value of 50 the corrections range from about 0.2% at $n = 1$ to 6% at $n = 7$. Highly anisotropic materials such as longitudinally reinforced fiber composites

Table 1.4   Added Mass Factors $\langle W_n^2(x)/W_n^2(0) \rangle$ for a Free–Free Vibrating Beam

$\dfrac{\Delta}{l}$	n						
	1	*2*	*3*	*4*	*5*	*6*	*7*
0.018	0.92	0.87	0.82	0.77	0.72	0.68	0.64
0.022	0.90	0.84	0.78	0.72	0.67	0.62	0.57
0.026	0.88	0.81	0.74	0.68	0.62	0.56	0.51
0.030	0.87	0.78	0.71	0.64	0.57	0.51	0.46
0.034	0.85	0.76	0.67	0.60	0.53	0.47	0.42
0.038	0.83	0.73	0.64	0.56	0.49	0.43	0.37
0.042	0.82	0.71	0.61	0.53	0.45	0.39	0.34

have relatively large values of E_1'/G_{13}' and require larger shear corrections. In such cases more accurate approximations than (51) can be employed [7].

Various expressions can be derived from the classical theory for the determination of the loss factor. It follows, for example, from equation (5.21) of [7] that

$$\tan \delta = \frac{f_{2n}^2 - f_{rn}^2}{f_{rn}^2} \qquad (52)$$

where f_{2n} is the frequency above f_{rn} at which the vibration amplitude is $1/\sqrt{2}$ times the resonance amplitude, corresponding to a 3-dB amplitude drop (Figure 1.13). If f_{1n} is the frequency below f_{rn} at which the amplitude is 3 dB below the maximum, then, with the approximation $f_{1n} + f_{2n} \approx 2f_{rn}$, it can be shown that

$$\tan \delta = \frac{f_{2n} - f_{1n}}{f_{rn}} \qquad (53)$$

Values of $\tan \delta$ up to about 0.4 can be obtained using (52) or (53) from fundamental resonance peaks since these peaks are then well resolved. The second and higher mode resonance peaks are much more influenced by overlap from the lower mode and larger magnitude peaks, and with increasing loss they become broadened on the low-frequency side. Since (52) involves measurements on the high-frequency side of the peak, it minimizes the effects of asymmetry; thus, it is routinely employed at the NPL for loss-factor determinations. For the higher modes, $\tan \delta$ values up to about 0.15 can be determined by this method.

The bandwidth method cannot be used to determine $\tan \delta$ values accurately below about 0.02 for $n = 1$ and below 0.002 for $n > 1$. In these cases measurements are made of the free decay of vibration amplitudes after cutting the drive force. If $W_n(t_1)$ and $W_n(t_2)$ are the amplitudes at the respective times t_1 and t_2 after removing the external force, then for small damping

$$\tan \delta = \frac{\ln[W_n(t_1)/W_n(t_2)]}{\pi f_{rn}(t_2 - t_1)} \qquad (54)$$

Values of loss factor are not generally influenced by added mass, and for isotropic materials corrections to $\tan \delta_E$ for shear deformations are usually negligible since $\tan \delta_G$ and $\tan \delta_E$ values are approximately equal. However, anisotropic materials with large E_1'/G_{13}' values and large differences between $\tan \delta_G$ and $\tan \delta_E$ may require large shear corrections to $\tan \delta_E$ [7].

In addition to the influence of added mass, shear, and rotatory inertia, other potential sources of error with the flexural resonance technique have been discussed in detail [7]. If the supporting loops are located within 0.5 mm of the nodes, then errors in E' that are caused by the added stiffness of the suspension are reduced to below 0.2% and errors in $\tan \delta$ that are caused by energy dissipation through the suspension are minimized. Residual contributions to $\tan \delta$ up to 0.002 may still be significant for low-loss materials, although empirical corrections can be applied. Contributions to $\tan \delta$ from air damping up to 0.001 are possible, but these are usually negligible or are corrected for when necessary. The presence of a static magnetic field on the drive transducer can significantly affect E' and $\tan \delta_E$ values obtained from measurements in the fundamental mode. This effect is related to the potential energy contribution from the field, and it is eliminated by ensuring that the spacing between the transducer and drive mass exceeds about 5 mm for $n = 1$.

Determinations of E' and $\tan \delta_E$ for flexible, high-loss polymers were made at audiofrequencies by flexural resonance tests on two-layer beams with the material bonded to a strip of metal or rigid plastic [7]. The relevant theory is complex, particularly when the effects of shear and rotatory inertia are included in the analysis of data for the higher modes. However, the analysis can be handled routinely with simple computer programs.

4.2 Torsional Resonance

Torsional vibrations can be induced in strips of rectangular cross section by bonding small pieces of magnetic steel at the corners of opposite faces at one end of the strip (Figure 1.16). An alternating torque is then produced by two drive

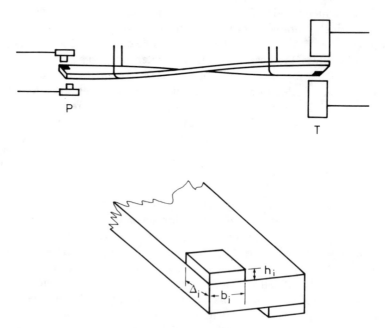

Figure 1.16 Schematic illustration of the torsional resonance method and dimensions of added drive and detector masses.

transducers T, which act on the respective steel masses. The transducers are fed from the power amplifier by separate current outputs that can be varied to balance the forces on the two masses. Clamping effects are best avoided by studying free–free torsional oscillations of samples suspended by fine nylon loops. Vibration amplitudes are then monitored with two proximitor probes P facing metallic masses bonded at the corners of the end of the specimen opposite to the drive transducers. The relative phase of the two detector outputs can distinguish between torsional and flexural modes.

Free torsional vibrations of long, thin, low-loss isotropic strips are governed by the equation of motion

$$\frac{KG^*}{\rho I_p} \frac{\partial^2 \theta}{\partial x^2} - \frac{\partial^2 \theta}{\partial t^2} = 0 \tag{55}$$

where θ is the angle of twist at time t of an element located a distance x along the beam, I_p is the polar moment of inertia, and KG^* is the torsional rigidity of a unit length of sample. For rectangular cross sections, $I_p = bh(b^2 + h^2)/12$ and $K = bh^3[1 - 0.63(h/b)]/3$. Solutions to (55) are of the form $\theta_n(x, t) = \Theta_n(x)\exp(i\omega_n^* t)$, where the natural frequencies ω_n^* and mode shapes $\Theta_n(x)$ are related to eigenvalues a_n determined by the sample boundary conditions. For free–free beams $a_n = n\pi$ and the mode shapes are sinusoidal (Figure 1.17).

When a harmonic external torque is applied to one end of the sample (Figure 1.16) the displacement amplitude–frequency relationship can be derived from the modified equation of motion [7]. For $\tan \delta_G$ below about 0.1 it follows that G' can be evaluated from the frequencies f_{rn} of the resolved resonances. With the inclusion of an added-mass correction factor $(1 + \psi_{mt})$ the relevant equation is

$$G' = \frac{4\rho I_p l^2 f_{rn}^2}{Kn^2} (1 + \psi_{mt}) \tag{56}$$

If the added masses are confined to sufficiently small regions at the end of the strip such that Δ_i/l does not exceed about 0.03 (Figure 1.16), then, for the lowest seven modes of a free–free beam ψ_{mt} is given to a good approximation by [7]:

$$\psi_{mt} = 2 \sum_i \frac{m_i}{M} \left[\frac{4(b_i^2 + h_i^2) - 6(bb_i - hh_i)}{b^2 + h^2} + 3 \right] \tag{57}$$

where M is the sample mass, m_i is the ith added mass ($i = 1$–4), and b_i and h_i are shown in Figure 1.16. According to (56) and (57) added masses around 1% typically yield corrections to G' up to 4%.

In addition to the influence of added mass, values of G' may be affected by axial stresses and inertial forces arising from the warping of rectangular cross sections. Corrections for these effects are rather involved but often negligible for the lowest three torsional modes [7].

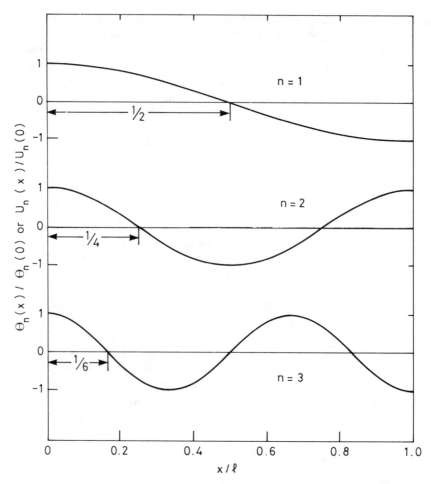

Figure 1.17   Theoretical natural-mode shapes for torsional or longitudinal vibrations of a free–free beam.

Values of $\tan \delta_G$ from torsional resonance can be obtained using (52)–(54), but remember the associated restrictions on their ranges of validity. Values of $\tan \delta_G$ are little affected by added mass or by warping effects in isotropic materials. Errors in G' and $\tan \delta_G$ that are caused by the supporting strings, static magnetic transducer fields, and air damping are usually negligible with the NPL setup.

4.3   Longitudinal Resonance

For the excitation and detection of longitudinal free–free vibrations, steel masses m_1 and m_2 are bonded at each end of the sample so as to face a magnetic drive transducer T at one end and a proximitor detector P at the other (Figure

1.18). The sample can be gripped by needle supports N to resist the static component of the magnetic field.

In the absence of an external force, the longitudinal vibrations can be analyzed approximately by the equation

$$\frac{E^*}{\rho} \frac{\partial^2 u}{\partial x^2} - \frac{\partial^2 u}{\partial t^2} = 0 \tag{58}$$

where u is the longitudinal displacement at time t of an element located a distance x along the beam. Solutions to (58) serve to define natural frequencies ω_n^* and mode shapes $U_n(x)$ related to eigenvalues c_n determined by the sample boundary conditions. Free–free vibrations have $c_n = n\pi$ and mode shapes of the sinusoidal form observed with free–free torsional modes (Figure 1.17). The addition of an external force term in (58) yields a relationship between the displacement amplitude and drive frequency that for sufficiently low loss describes the resonance peaks shown schematically in Figure 1.13. It follows that E' can be determined from

$$E' = \frac{4\rho l^2 f_{rn}^2}{n^2} \left[1 + 2\frac{(m_1 + m_2)}{M} \right] \left[1 + \frac{n^2\pi^2 v^2}{12} \frac{(b^2 + h^2)}{l^2} \right] \tag{59}$$

The second term on the right-hand side of (59) represents a correction for the added masses. The third term is a small correction factor that allows for the inertial effects of lateral motions of beam elements and depends on Poisson's ratio v.

Loss factors $\tan \delta_E$ from longitudinal resonance are obtained using (52)–(54). Because of the relatively high frequencies or effective sample stiffnesses associated with the longitudinal resonance technique, the effects on E' and $\tan \delta_E$ caused by the supports, the static transducer field, and air damping are usually negligible [7].

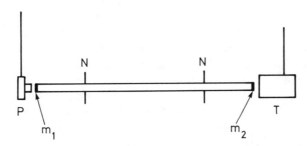

Figure 1.18 Illustration of longitudinal resonance technique and added drive and detector masses.

4.4   Illustration of Results for Poly(propylene)

Results shown in Figure 1.19 for poly(propylene) provide a good example of the value of the audiofrequency resonance technique since the glass–rubber relaxation region for this material (usually labeled β) is centered in the audiofrequency range at room temperature. This diagram, which includes plots of E' and $\tan \delta_E$ versus frequency over the range from 10^{-2} to $10^{6.4}$ Hz at temperatures between -10 and $+50°C$, also exemplifies the agreement between the flexural and longitudinal resonance data and results obtained by other methods.

An Imperial Chemical Industries grade PXC 8830 poly(propylene) that had been annealed at $130°C$ and slowly cooled to room temperature was obtained in strip form from Dr. McCrum of Oxford University. Measurements were made after the polymer had been stored at room temperature for about 6 months when the measured polymer density was 906.5 kg/m^3. To minimize subsequent deaging effects the measurements at 30, 40, and $50°C$ were made after completion of the lower temperature studies. At each temperature the sample was held for about 1 h before making measurements.

In the frequency range of 10^{-2}–10^2 Hz the results were obtained with the tensile nonresonance technique (Section 2) at a strain amplitude of 2×10^{-4}. At frequencies between 160 Hz and 4.4 kHz the data were determined from flexural resonance measurements on modes 1–5 for a sample with $l/h = 49.9$. Corrections to E' for added mass varied from 3.9 to 2.5% and shear and rotatory inertia corrections ranged from 0.3 to 4.2% as n increased from 1 to 5. Measurements on the fundamental longitudinal resonance mode yielded data between 5 and 10 kHz, the corrections to E' for added mass and lateral inertia amounting to 2.1 and 0.08%, respectively. At ultrasonic frequencies E' and $\tan \delta_E$ were evaluated from the components of L^* and G^* on the basis of (14). These components were determined as described in Section 5 from the velocity and attenuation of pulses with mean frequencies of 1.1 and 2.8 MHz. Data obtained by the various methods lie on quite smooth curves (Figure 1.19) consistent with an overall accuracy of around 2% in E' and 5% in $\tan \delta_E$.

The most notable feature in Figure 1.19 is the increase in magnitude of the β relaxation with increasing temperature as is evident from the increased area under the loss peak. This result could reflect an increased population of high-energy molecular conformations in the amorphous phase [28], an effect that could relate to some deaging of the structure on heating to temperatures above about $30°C$. The breadth of the $\tan \delta_E$ peak is substantially larger than that (1.14 decades) for a single-relaxation-time model and increases with decreasing temperature. The theoretical curve in Figure 1.20 illustrates that the temperature dependence of the $\tan \delta_E$ peak frequency ($f_{\text{max}} = 1/2\pi\bar{\tau}$) is non-Arrhenius, and it corresponds to the Vogel–Fulcher equation (20) with $\tau_0 = 10^{-12.1}$ s, $A = 1082$, and $T_\infty = 233$ K. Note that these results are consistent with the glass-transition assignment for the β process and do not support the predictions based on results from thermally stimulated creep (or so-called *thermal sampling*) investigations [55].

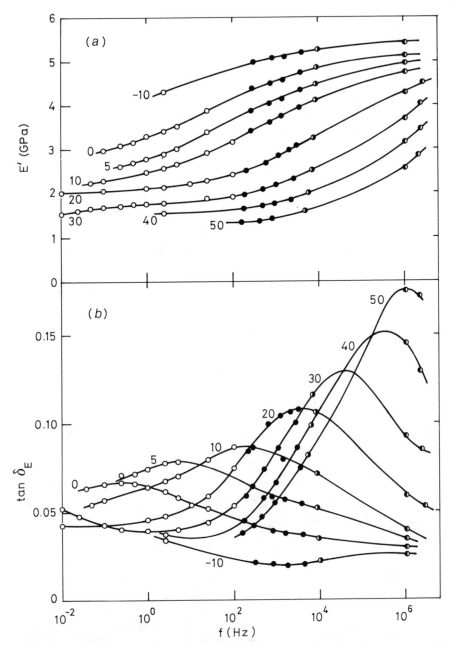

Figure 1.19 Frequency dependence of (*a*) E' and (*b*) tan δ_E for poly(propylene) at temperatures (°C) indicated on each curve. Symbols are ○, tensile nonresonance method; ●, flexural resonance; ◐, longitudinal resonance; and ◑, ultrasonic pulse technique.

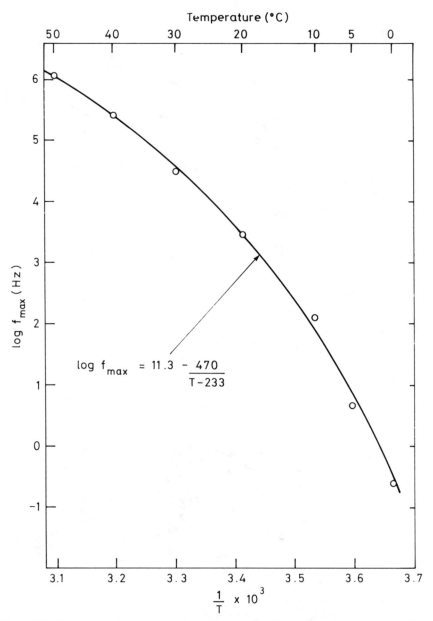

Figure 1.20 Dependence of log f_{max} on reciprocal temperature for poly(propylene), where f_{max} is the frequency of maximum tan δ in constant temperature measurements. The experimental points (O) are taken from Figure 1.19. The theoretical curve corresponds to the Vogel–Fulcher equation (20) as discussed in the text.

5   ULTRASONIC WAVE TECHNIQUES

5.1   Introduction

Ultrasonic wave propagation techniques enable measurements to be made of the phase velocity and attenuation of acoustic waves traveling along known directions in a material. From these quantities, the storage and loss components of complex moduli can be determined. Test methods operate in the frequency range from below 100 kHz to over 100 MHz and so complement other methods by extending measurements to higher frequencies. They are particularly attractive for the measurement of elastic anisotropy in single crystals, fiber-reinforced plastics, and oriented thermoplastics and for the study of relaxation mechanisms in thermoplastics, liquids, and solutions. Measurements are made rapidly and, generally, with very high accuracy.

A sound wave traveling in a material whose boundaries are sufficiently remote that they do not influence the mode of propagation is termed a *bulk wave*. The requirement for remote boundaries is satisfied if the sound wavelength is somewhat smaller than the test-piece dimensions in the plane normal to the direction of propagation. In isotropic elastic solids there are two types of bulk wave that can be propagated, which are characterized by the polarization of the displacement vector. The displacement is thus directed along (longitudinal wave) or normal to (transverse wave) the direction of wave propagation. The relationship between wave velocities and elastic properties is then very simple (see Section 5.2).

In anisotropic materials, there are three types of bulk wave that can be propagated, and in an arbitrary direction their polarizations are quasilongitudinal or quasitransverse. The relationships between wave velocity, material properties, and wave direction then contain more than one elastic constant. Along symmetry directions, however, these relationships are analogous to those for isotropic materials, the appropriate material property being dependent on the symmetry direction. In principle, by measurement of wave velocities in a sufficient number of directions, all the elastic moduli of a material of arbitrary symmetry can be obtained. In practice, the symmetry of materials such as fiber-reinforced plastics and oriented polymers can be assumed to be hexagonal, and a very much reduced set of measurements is sufficient. Additional measurements then serve to identify symmetry axes, such as directions of preferred fiber or molecular orientation.

The most direct method used to determine velocity and attenuation involves measurement of the time of flight and the reduction in amplitude of an acoustic pulse traveling a known path length in a test piece. This method is suitable for isotropic and anisotropic solids in the frequency range from 0.1 to 10 MHz and is the principle behind the measurement techniques and analyses described here. At frequencies above about 10 MHz, the attenuation, which increases with frequency, is usually so high that only small path lengths can be adopted and the accuracy in measurements is correspondingly low. A number of other methods are then more appropriate and are applicable to films and liquids [13, 56]. These can operate over a wide frequency range. Techniques employing acoustic interferometers [57, 58] and the measurement of complex impedance [13, 59] are particularly suited to liquids.

In those materials where the velocity or attenuation is dependent on frequency in the frequency range spanned by the ultrasonic pulse, the shape of the pulse changes on transmission through a test piece. Conventional methods described in Section 5.3.2 for determining pulse transit time or attenuation are then subject to some inaccuracy, which can be avoided by using a Fourier transform method as explained in Section 5.3.3. Where the source of pulse distortion stems from the viscoelastic behavior of the material, the Fourier method enables storage moduli and loss factors to be rigorously determined over the frequency range spanned by the pulse.

Whereas bulk waves travel remote from the materials surface, there are other waves that can be propagated whose character is determined by stress-free boundary conditions. These waves travel parallel to, and in the vicinity of, the surface. A number of modes can be generated depending on the wavelength in comparison with the thickness of a test piece. In the limit of short wavelength (or large thickness), the surface or Rayleigh wave only is propagated, and the wave amplitude decays exponentially with distance from the single surface. When the wavelength is comparable with the test-piece thickness, the opposing stress-free surface influences the mode of propagation and the generation of a series of plate modes is possible. These are also termed *Lamb waves*.

The use of surface or plate waves for materials characterization is potentially attractive because they can be launched and received from a single surface. The attenuation is lower than that for bulk waves of similar frequency so large path lengths can be employed and the accuracy in measurements is correspondingly high. However, relationships between wave velocity and materials properties are complicated, especially in anisotropic materials [60], and contain more than one modulus. Therefore, investigations employing these waves are concerned mainly with nondestructive testing such as the detection of cracks or frozen-in stress [61–64].

5.2   Theory for Bulk Wave Propagation

5.2.1   Anisotropic Elastic Materials

For elastic materials the loss factors are negligible and the low-strain mechanical properties are specified in terms of elastic moduli, which become identical to the storage moduli. When such materials are also anisotropic, their behavior can be characterized by the components of a stiffness matrix c_{ij} ($i, j = 1$–6) [6, 7], which are related to modulus components such as E_1, E_2, and G_{12} (Section 1.2) by (95) (Section 5.4.1).

The theory of acoustic wave propagation in anisotropic materials has been developed by Musgrave [65]. Its application to materials of hexagonal symmetry is considered in [7, 66] because, to a good approximation, this symmetry class is appropriate for the majority of anisotropic plastics, including fiber-reinforced materials. There are then five independent components of the stiffness matrix; and, assuming the 1 axis is the unique axis, these are $c_{11}, c_{22} = c_{33}; c_{44}, c_{55} = c_{66}; c_{12} = c_{13}$; and $c_{23} = c_{22} - 2c_{44}$. An arbitrary direction in the material can be defined by the angle θ between this direction and the unique axis. The 2 and 3 axes can now be chosen such that the 3 axis (say) is normal to the plane containing the arbitrary direction and the unique axis. The arbitrary direction is then defined by the unit vector $(n_1, n_2, 0)$, where $n_1 = \cos\theta$ and $n_2 = \sin\theta$. The velocities v of the three bulk waves that are propagated with wave normal along $(n_1, n_2, 0)$ are given by the roots of the cubic equation [7, 66]

$$(n_1^2 c_{66} + n_2^2 c_{44} - \rho v^2)$$
$$\times [(n_1^2 c_{66} + n_2^2 c_{22} - \rho v^2)(n_1^2 c_{11} + n_2^2 c_{66} - \rho v^2) - n_1^2 n_2^2 (c_{12} + c_{66})^2] = 0$$

$$(60)$$

The root

$$\rho v^2 = n_1^2 c_{66} + n_2^2 c_{44} \tag{61}$$

gives the velocity of a pure transverse wave that is polarized normal to the 1–2 plane. The roots of the quadratic factor

$$(n_1^2 c_{66} + n_2^2 c_{22} - \rho v^2)(n_1^2 c_{11} + n_2^2 c_{66} - \rho v^2) - n_1^2 n_2^2 (c_{12} + c_{66})^2 = 0 \tag{62}$$

give the velocities of two waves that are polarized in the 1–2 plane; the faster is quasilongitudinal and the other is quasitransverse. For propagation along the unique axis, $n_1 = 1$, $n_2 = 0$, and the roots of (62) become

$$v = (c_{11}/\rho)^{1/2} \tag{63}$$

and

$$v = (c_{66}/\rho)^{1/2} \tag{64}$$

The first of these roots now corresponds to a pure longitudinal wave and the second to a pure transverse wave. Perpendicular to the unique axis, the wave normal is $(0, 1, 0)$, and (62) gives

$$v = (c_{22}/\rho)^{1/2} \tag{65}$$

and

$$v = (c_{66}/\rho)^{1/2} \tag{66}$$

for the velocities of the pure longitudinal and pure transverse waves, respectively, in this direction.

Propagation in a plane normal to the unique axis is characterized by the equation [7, 66]

$$(\rho v^2 - c_{44})(\rho v^2 - c_{22})(\rho v^2 - c_{66}) = 0 \tag{67}$$

for which the roots are independent of direction as expected in a plane of isotropy. The roots $v = (c_{66}/\rho)^{1/2}$ and $v = (c_{44}/\rho)^{1/2}$ correspond to pure shear waves polarized out of, and in, the plane of isotropy, respectively. The wave with velocity $v = (c_{22}/\rho)^{1/2}$ is pure longitudinal.

It is apparent that measurement of the velocities of longitudinal and shear waves traveling along symmetry directions in a material enable four of the five stiffness components to be determined. Measurement of wave velocity in the 1–2 plane enables the fifth component c_{12} to be obtained from (62). Apparatus for carrying out these measurements is described in Section 5.3.1.

5.2.2   Isotropic Elastic Materials

For isotropic materials

$$c_{11} = c_{22} = L$$
$$c_{44} = c_{66} = G \tag{68}$$

and

$$c_{12} = c_{23} = L - 2G$$

Therefore (60) has roots

$$v_L = (L/\rho)^{1/2} \tag{69}$$

and

$$v_T = (G/\rho)^{1/2} \tag{70}$$

representing the velocities of the two bulk waves, longitudinal and transverse, respectively, that can be generated in every direction.

5.2.3   Viscoelastic Materials

With viscoelastic materials, stiffness parameters become complex quantities. Considering isotropic materials for the moment, (69) demonstrates that a complex longitudinal stiffness, $L^* = L' + iL''$, gives rise to a complex longitudinal wave velocity, $v_L^* = v_L' + iv_L''$, such that

$$L' + iL'' = \rho(v_L' + iv_L'')^2 \tag{71}$$

Under these circumstances, the wave displacement is [7, 66]

$$u_L = A_L \exp i\omega \left[t - x \frac{v'_L}{v'^2_L + v''^2_L} \right] \exp \left[\frac{-\omega v''_L x}{(v'^2_L + v''^2_L)} \right] \tag{72}$$

This represents a harmonic wave whose amplitude is attenuated exponentially with distance x in the direction of propagation. The parameter A_L is the amplitude of the wave at the point where it is launched. The phase velocity v_L is now

$$v_L = \frac{v'^2_L + v''^2_L}{v'_L} \tag{73}$$

and the absorption coefficient α_L is

$$\alpha_L = \frac{\omega v''_L}{(v'^2_L + v''^2_L)} \tag{74}$$

Combining (71), (73), and (74) we have

$$L' = \frac{\rho v^2_L \omega^2 (\omega^2 - \alpha^2_L v^2_L)}{(\omega^2 + \alpha^2_L v^2_L)^2}$$

and   (75)

$$L'' = \frac{2\rho \alpha_L v^3_L \omega^3}{(\omega^2 + \alpha^2_L v^2_L)^2}$$

Therefore, the loss tangent $\tan \delta_L$ is

$$\tan \delta_L = \frac{2\alpha_L v_L \omega}{\omega^2 - \alpha^2_L v^2_L} \tag{76}$$

On expressing (76) in terms of α_L, it is apparent that for materials with $\tan \delta_L \leqslant 0.2$,

$$\tan \delta_L = \frac{2\alpha_L v_L}{\omega} \tag{77}$$

with an error of less than 1%. The implication for such materials is that $\omega^2 \gg \alpha^2 v^2_L$, and so

$$L' = \rho v^2_L \tag{78}$$

The shear loss factor $\tan \delta_G$ and the storage shear modulus G' are similarly related to the absorption coefficient and velocity of the transverse wave. Thus

$$G' = \rho v_T^2 \qquad (79)$$

$$\tan \delta_G = \frac{2\alpha_T v_T}{\omega} \qquad (80)$$

From measurements of L', G', $\tan \delta_L$, and $\tan \delta_G$, the Young's storage modulus E' and loss factor $\tan \delta_E$ are obtained from the real and imaginary components of (14). Thus

$$E' = \frac{G'(3L' - 4G')}{L' - G'} \qquad (81)$$

$$\tan \delta_E = \tan \delta_G - \frac{G'L'(\tan \delta_G - \tan \delta_L)}{(L' - G')(3L' - 4G')} \qquad (82)$$

Likewise the bulk storage modulus and loss factor can be calculated from the components of (13).

Along symmetry directions in anisotropic viscoelastic materials, analogous relationships to (77)–(80) exist for complex tensile and shear stiffness components. Relationships for the components of the complex stiffness c_{12}^* are more complicated, and there is unlikely to be experimental interest in the imaginary component c_{12}'' because of the high accuracy needed in associated measurements to achieve moderate accuracy in c_{12}''.

5.3   Measurement of Bulk Wave Velocity and Attenuation

5.3.1   Apparatus

The most direct method of velocity measurement consists of timing the transit of an ultrasonic pulse through a known path length in a test piece [7, 66–68]. A convenient apparatus for launching the wave is depicted in Figure 1.21. It consists of two coaxial ultrasonic transducers immersed in a water bath whose temperature is accurately controlled to within $\pm 0.1°C$. One acts as a transmitter, the other as a receiver. For isotropic materials, the test piece requires only one pair of parallel faces. For anisotropic samples, measurements along a larger number of directions are needed and the test piece should take the form of a rectangular parallelepiped. The test piece is mounted vertically on a turntable between the transducers to intercept the beam. The angle of incidence can be varied by a stepper motor and set to $\pm 0.1°$. At normal incidence, only longitudinal waves are generated in the test piece. At arbitrary angles, two refracted waves can be launched; one is quasilongitudinal (pure longitudinal in isotropic materials), the other is quasitransverse (pure transverse in the isotropic case). These waves are polarized in the plane containing the incident and refracted wave directions. For reliable measurements of absorption coefficient

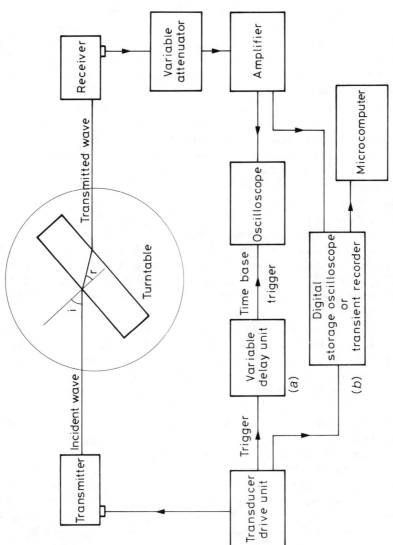

Figure 1.21 The ultrasonic pulse apparatus showing electronic components required for (a) the direct measurement of pulse arrival time and (b) a Fourier transform method for pulse analysis.

55

using refracted waves, it is necessary to select a receiving transducer whose diameter is greater than the beam diameter from the transmitter.

An alternative experimental arrangement, suitable mainly for isotropic materials, allows direct contact between the transducers and the faces of the test piece. The water bath and turntable are then not needed. The transducer separation can be varied to accommodate samples of different thickness including zero thickness. Both longitudinal and shear wave transducer pairs are required. This arrangement is not suitable for determining all the stiffness components of anisotropic materials because of the requirement then to launch waves in a range of directions. It is also susceptible to a small error arising from the need to introduce a coupling medium between the transducers and the test piece. Furthermore, the arrival time and, to a greater extent, the received amplitude are sensitive to the contact pressure.

The transmitting transducer is powered by a series of high-voltage, short-duration pulses from the transducer drive unit (typically, 300 V for 10 ns). A pulse repetition interval of about 1 ms is satisfactory. The pulses are detected by the receiving transducer and amplified by a circuit with an accurate attenuator. An example of a pulse waveform is given in Figure 1.22, the frequency and length being determined by the transducer design.

Commerical apparatus similar to that shown in Figure 1.21 is not available from a single manufacturer. Several companies, such as Panametrics Inc. in the United States (the United Kingdom agent is Diagnostic Sonar Ltd.) and Wells Krautkramer in the United Kingom and the Federal Republic of Germany, manufacture ultrasonic probes, transducer drive units, and amplifiers. Measurements of arrival time can be made using an oscilloscope whose specification is sufficiently high that its variable time-base trigger delay has nanosecond accuracy and sensitivity. Similarly, several top-of-the-range digital storage oscilloscopes have sufficient sampling speed to record the pulse, and a number of these have a peripheral microcomputer that can be programmed to carry out a Fourier transform as a standard package.

The Matec Company (United Kingdom agent is The Roditi International Corporation Ltd.) markets a complete experimental arrangement, less the water bath and turntable, that determines wave velocity and absorption coefficient by a different principle from that depicted in Figure 1.21. The transmitting transducer is powered by a chopped, sinusoidal signal whose frequency can be varied. A reference waveform from the transducer drive unit is used to determine the phase of the received waveform. By measuring phase values corresponding to different drive frequencies, the phase velocity can be deduced [69]. Corrections need to be applied to take account of certain sources of extraneous phase. When these corrections are applied, uncertainties in transit time measurements of typically a few nanoseconds are reported [69], which is comparable to the accuracy achievable with the method described here.

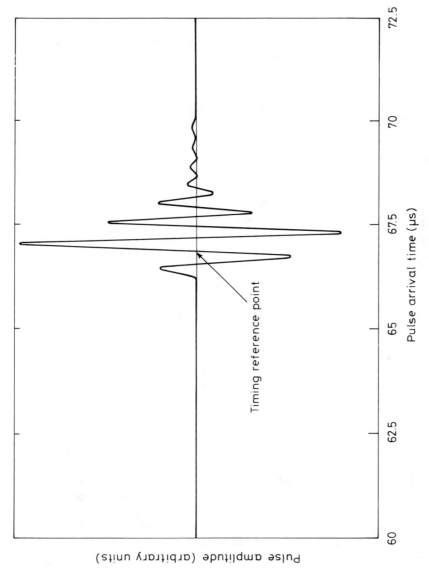

Figure 1.22 Oscilloscope trace of the waveform of a received pulse through water as the immersion medium. The nominal frequency of the transducer is 2 MHz, and a time-base delay of 60 μs was selected.

Pulse amplitude (arbitrary units)

Pulse arrival time (μs)

Timing reference point

5.3.2 Pulse Comparison Method

The amplified pulse waveform from the receiver transducer is displayed on an oscilloscope whose time base is triggered by the transmitter pulse generator through an accurate, variable delay unit (see Figure 1.21a). This unit is capable of inserting delay times greater than the pulse transit time between the transducers, and it has a resolution in the region of 1 ns. Thus, the arrival time of the pulse can be determined by adjusting the delay until some characteristic feature of the pulse coincides with the center of the graticule on the oscilloscope screen. An early, zero-voltage crossover is typically used as the timing feature, as indicated in Figure 1.22. When a test piece is inserted in the beam, a new arrival time can be determined. From measurements of the difference τ between the arrival times through water and the sample and the angle of incidence i, the pulse velocity v can be calculated from

$$\tau = \frac{b}{\cos r}\left[\frac{\cos(r-i)}{v_W} - \frac{1}{v}\right] \tag{83}$$

where v_W is the velocity of sound in water, b is the specimen dimension normal to the incidence face, and r is the angle of refraction. The latter is given by

$$\frac{\sin r}{v} = \frac{\sin i}{v_W} \tag{84}$$

yielding, from (83)

$$\tan r = \frac{\sin i}{\cos i - v_W \tau/b} \tag{85}$$

Longitudinal wave velocities v_L are more conveniently obtained at normal incidence where (83) gives

$$v_L = \frac{b v_W}{b - \tau v_W} \tag{86}$$

The absorption coefficient can be determined from the change in attenuator setting needed to bring the pulse amplitude back to the level obtained in the absence of the test piece. This attenuation is a combination of losses arising from absorption and reflection. The reflection loss can be eliminated through attenuation measurements at a constant angle of incidence on samples of more than one thickness. As shown in [7], if a_1 and a_2 correspond to attenuator settings in decibels for samples of thickness b_1 and b_2, then the absorption coefficient α is given by

$$\alpha = \frac{0.115(a_1 - a_2)}{(b_2 - b_1)\sec r} \tag{87}$$

5.3.3   Fourier Transform Method

Relaxation mechanisms in viscoelastic materials give rise to properties that vary with time or frequency. Where the distribution of relaxation frequencies for a mechanism extends into the ultrasonic frequency range, the wave velocity and attenuation are frequency dependent. This leads to a change in the shape of a pulse transmitted by the test piece; the shorter is the pulse, the greater the change. Scattering by inclusions such as fibers or voids is another factor that leads to pulse distortion. Figure 1.23 illustrates the changes in pulse shape that can arise because of a frequency dependence of properties. Waveform part (b)

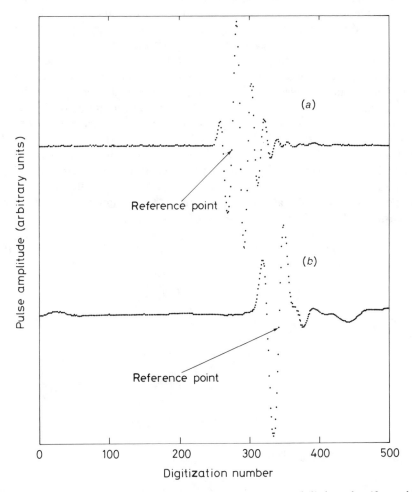

Figure 1.23 Digital record of pulse waveforms (a) through water and (b) through a 12-mm-thick poly(propylene) test piece at a 45° angle of incidence. The sampling interval is 25 ns, and the difference in attenuator setting is 36 dB.

represents the transverse wave generated by the water pulse shown in part (a) incident at an angle of 45° on a 12-mm-thick poly(propylene) sample. Under these circumstances, measurements of changes in velocity or amplitude by the pulse comparison method described in Section 5.3.2 are subject to an uncertain error. For higher accuracy, it is necessary to carry out Fourier transforms of the received pulses to derive information on the phase and amplitude of discrete frequency components. This requires obtaining digital records of each pulse and performing the Fourier transforms by microcomputer [70, 71].

The additional apparatus needed is shown in Figure 1.21b. The oscilloscope of Figure 1.21a is replaced by a digital storage oscilloscope with a sampling interval substantially shorter than the mean period of the pulse. The high-resolution delay circuit is not required as long as the storage oscilloscope has a coarse trigger delay facility to enable the start of the digital records to be chosen fairly close to the beginning of each pulse. The microcomputer is programmed to carry out both the sine and cosine fast Fourier transforms.

For details of the theory behind the fast Fourier transform, the reader is referred to appropriate textbooks. An outline of the mathematical procedure is given here to aid the selection of suitable apparatus and operating conditions. The transient recorder in the oscilloscope records the amplitudes of a waveform at a selected sampling interval Δt. These are stored as a series of N discrete voltages $V(t_q)$ at time intervals $t_q = q\Delta t$ after the unit is triggered externally by the transducer drive unit following a suitable delay. The parameter q varies integrally between 0 and $N - 1$. The fast Fourier transform requires N to equal an integral power of 2. The amplitudes $A(f_p)$ and the phases $\varphi(f_p)$ of the components of the pulse at discrete frequencies f_p can now be calculated from the cosine and sine Fourier transforms, $C(f_p)$ and $S(f_p)$, respectively, using

$$A(f_p) = [C^2(f_p) + S^2(f_p)]^{1/2} \tag{88}$$

$$\varphi(f_p) = \tan^{-1}\left[\frac{S(f_p)}{C(f_p)}\right] \tag{89}$$

where

$$C(f_p) = \sum_{q=0}^{N-1} V(t_q) \cos 2\pi f_p q\Delta t \tag{90}$$

and

$$S(f_p) = \sum_{q=0}^{N-1} V(t_q) \sin 2\pi f_p q\Delta t \tag{91}$$

The allowable discrete frequencies f_p are given by

$$f_p = \frac{p}{N\Delta t} \tag{92}$$

where p is $1 \cdots N/2$.

The transit time at any frequency f_p is related to the phase difference at f_p between pulses with and without the test piece in the beam. However, the phase as given by (89) always lies between $-\pi$ and π. Phase differences greater than π are therefore uncertain by an integral number of multiples of 2π. To resolve this, it is necessary to shift the recorded window of the sample waveform such that the pulse occupies approximately the same position within the window as does the pulse through water alone. Denoting this time shift by τ_s and the respective phase values for the water and shifted pulses by $\varphi(f_p)$ and $\varphi_s(f_p)$, then the transit time τ at the frequency f_p is given by

$$\tau(f_p) = \tau_s + \frac{\varphi_s(f_p) - \varphi(f_p)}{2\pi f_p} \qquad (93)$$

The amplitudes $A(f_p)$ of the frequency components of each pulse in Figure 1.23 are plotted at a reduced set of frequencies around 2 MHz in Figure 1.24. The enhanced attenuation of the higher frequency components of the transverse wave is apparent. The low amplitudes of frequency components below about 400 kHz and above about 3 MHz restricts the derivation of properties to within this range. This usable frequency range can be extended by using a drive transducer that generates a somewhat shorter pulse. If the heights of the spectra of the transverse wave through samples of thickness b_1 and b_2 are $A_1(f_p)$ and $A_2(f_p)$, respectively, and a is the difference in attenuator settings, then the absorption coefficient is now

$$\alpha(f_p) = \frac{1}{(b_2 - b_1)\sec r}\left[0.115a + \ln \frac{A_1(f_p)}{A_2(f_p)}\right] \qquad (94)$$

5.3.4   Data Acquisition for the Fourier Transform

Equation (92) shows that the maximum frequency at which properties may be evaluated is $1/2\Delta t$. This frequency should be chosen so that it is above the frequency range spanned by the pulse. This places an upper limit on the sampling interval Δt. The number of data values N can then be selected so that the whole of the pulse is recorded. However, there are certain advantages in reducing the sampling interval and increasing N despite the associated increase in computing time. First a decrease in Δt gives an increase in timing accuracy since the digital record will start an arbitrary time up to one sampling interval after the trigger from the transducer drive unit. This leads to an error of up to $\pm\Delta t$ in the transit time difference τ. Second, at shorter sampling intervals, there are more data characterizing the pulse, which results in improvements in the accuracy with which amplitude and phase values are determined through a reduction in the influence of noise on measurements of $V(t_q)$. Significant improvements in timing accuracy and S/N ratio can also be achieved through averaging several records of the pulse.

Figure 1.23a shows a digital record of a pulse obtained using 2-MHz transducers in the absence of a test piece. Waveform part (b) represents a transverse wave generated in a 12-mm-thick sample of poly(propylene) at an angle of incidence of 45°. A sampling interval of 25 ns was chosen and the average of 32 sweeps was recorded. A trigger delay of 60 μs was selected to bring the pulse within the sweep time of the recorder. By keeping this delay constant for any set of measurements, errors associated with the accuracy in the quoted magnitude of the delay are avoided.

After transferring a record to the microcomputer, a reference point is identified by the operator. This typically corresponds to an early zero crossover as indicated in Figure 1.23. The difference in reference point times obtained with and without a sample gives the quantity τ_s in (93). Truncation points can also be selected on either side of the pulse to eliminate any spurious pulses present and to keep the transform length to a minimum. The microcomputer then pads out one end of the waveform so that the reference point lies at the center and then pads out both ends equally until a total of 2^k points (k integral) are obtained.

5.4   Some Illustrative Results

The application of bulk wave velocity measurements to material property determinations is now considered for two classes of material. The first is a uniaxially aligned, carbon fiber reinforced sample of PEEK. Measurements of longitudinal wave velocity with direction in a plane normal to the fiber axis revealed that this material is accurately transversely isotropic. Pulse distortion was very small, which implies that void content and dispersion by viscoelastic phenomena were both low. Therefore, wave velocities were determined by the direct measurement of pulse arrival times.

The second material is an unfilled poly(propylene) homopolymer. The high-frequency region of the β-relaxation mechanism extends to ultrasonic frequencies at room temperature, which leads to pulse distortion by dispersion, as shown in Figure 1.23. Therefore, pulse waveforms are analyzed by the Fourier transform method, and storage moduli and loss factors are determined.

5.4.1   An Anisotropic Elastic Material

Sheets of carbon fiber reinforced PEEK were supplied by ICI. The fibers were type XAS, manufactured by Courtaulds Ltd., and occupied a volume fraction of 54%. A rectangular test piece was machined with dimensions $70 \times 15 \times 5.4$ mm, where the long dimension coincided with the direction of fiber alignment (1 axis). Tensile stiffnesses c_{11} and c_{22} were determined from the velocities of longitudinal waves traveling in directions along and normal to the fiber axis using (63) and (65). The transverse shear stiffness c_{44} was obtained from the velocity of the transverse wave generated by refraction in a plane of isotropy. This is accomplished by mounting the sample such that the fiber axis is perpendicular to the surface of the turntable (see Figure 1.21) and selecting suitable angles of

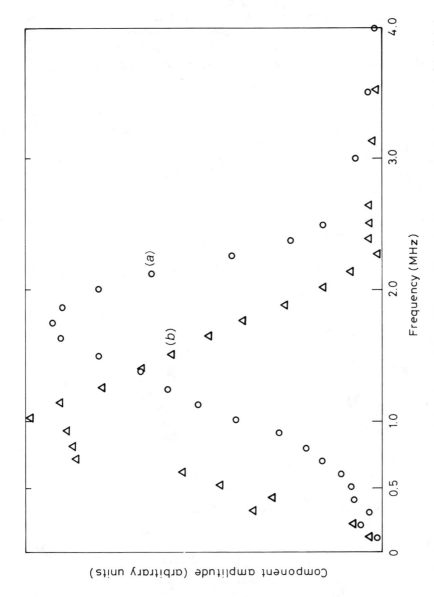

Figure 1.24 Spectral analyses of the pulse waveforms of Figure 1.23 obtained by taking Fourier transforms of the digital records. The difference in the attenuator settings is 36 dB.

63

incidence for generating the shear wave. The shear wave velocity is obtained using (84) and (85).

The remaining components c_{66} and c_{12} can be determined by measuring quasilongitudinal or quasitransverse wave velocities in a plane containing the fiber axis. By selecting a range of angles of incidence, values of velocity v with direction r can be obtained from (84) and (85). Substitution into (62) yields a series of equations that can be solved by an optimization routine to obtain the quantities c_{66} and c_{12}. For highly anisotropic materials where c_{11} is significantly greater than other stiffness components, expressions for the roots of (62) can be simplified [7] such that a plot of v_{qT}^2 versus $\cos^2 \theta$ is linear, where v_{qT} is the velocity of the quasitransverse wave. Components c_{66} and c_{12} are then obtained from the intercept and gradient of the line.

Table 1.5 gives measured values for the stiffness components and results for other, more familiar elastic properties (see Section 1.2) calculated from these using the following relationships:

$$E_1 = \frac{c_{11}(c_{22} + c_{23}) - 2c_{12}^2}{c_{22} + c_{12}}$$

$$E_2 = \frac{(c_{22} - c_{23})(c_{11}c_{22} + c_{23}c_{11} - 2c_{12}^2)}{c_{11}c_{22} - c_{12}^2} \qquad (95)$$

$$G_{12} = c_{66}$$

$$G_{23} = c_{44}$$

$$v_{21} = \frac{c_{12}}{c_{22} + c_{23}}$$

$$v_{23} = \frac{c_{23}c_{11} - c_{12}^2}{c_{22}c_{11} - c_{12}^2}$$

$$v_{12} = \frac{c_{12}(c_{22} - c_{23})}{c_{22}c_{11} - c_{12}^2}$$

Table 1.5 Elastic Properties of a Carbon Fiber Reinforced PEEK Material[a]

c_{11}	$c_{22} = c_{33}$	c_{44}	$c_{55} = c_{66}$	$c_{12} = c_{13}$	c_{23}	ρ
125	13.5	3.3	5.4	7.0	6.8	1586
E_1	$E_2 = E_3$	G_{23}	$G_{12} = G_{13}$	v_{21}	v_{23}	v_{12}
120	10.0	3.3	5.4	0.34	0.49	0.03

[a]The 1 axis is in the direction of fiber alignment. Units are gigapascals (GPa) for moduli and kilograms per cubic meter (kg/m³) for density.

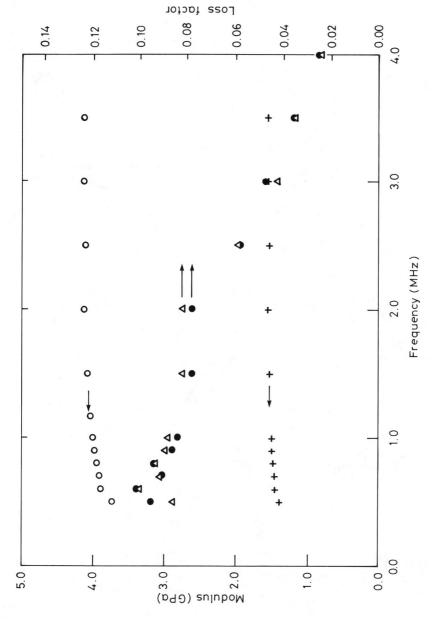

Figure 1.25 Frequency dependence of storage moduli and loss factors for poly(propylene) obtained by the Fourier transform method. Symbols are $\bigcirc$, E'; $+$, G'; $\bullet$, $\tan\delta_E$; and $\triangle$, $\tan\delta_G$.

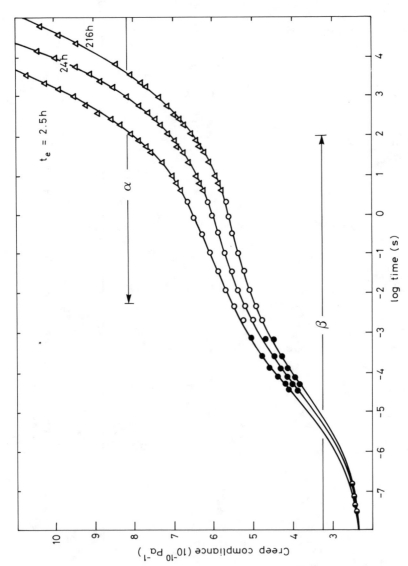

Figure 1.26 Comparison of tensile creep compliance data $D(t)$ for poly(propylene) derived using a combination of dynamic mechanical and static loading test methods: ◑, ultrasonics; ●, flexural resonance; ○, tensile nonresonance; and △, conventional creep. The three curves correspond to different states of physical aging and demonstrate the influence of physical aging on both the α- and β-relaxation mechanisms.

Poisson's ratios v_{ab} are defined as the ratio of the lateral strain along the a axis to the longitudinal strain along the b axis for uniaxial loading along the b axis.

5.4.2   An Isotropic Viscoelastic Material

Two compression-molded sheets of poly(propylene) were supplied by ICI of nominally 3 and 12 mm thickness. The density was $912\,kg/m^3$ for each sheet. Test pieces of approximately $70 \times 50\,mm$ were cut from regions of each sheet where the thickness was uniform. Longitudinal and shear wave velocities were measured over a range of frequencies f_p in the thicker test piece using the Fourier transform procedure and (84), (85), and (94). Young's and shear moduli at selected frequencies spanned by the pulse (see Figure 1.24) are presented in Figure 1.25. Absorption coefficients $\alpha_L(f_p)$ and $\alpha_T(f_p)$ were calculated using (88) and (94). Loss factors $\tan\delta_E$ and $\tan\delta_G$ were derived from (77), (80), and (82) and are also displayed in Figure 1.25.

From the ultrasonic E' and E'' data for poly(propylene), creep compliances $D(t)$ were evaluated using (16) to obtain D' followed by the frequency–time transformation (18). These results are compared in Figure 1.26 with similarly transformed dynamic data from audiofrequency and forced nonresonance measurements and with compliance values obtained from conventional creep experiments [18]. Results are shown at three states of physical aging characterized by the time t_e between rapid cooling from a temperature of 80°C and the start of the compliance measurements. They demonstrate how creep data from $10^{-8}\,s$ and spanning more than 13 decades of time can be obtained through the use of a combination of dynamic and static test methods. A comparison of measurements in the frequency or time regions where data from different methods overlap serves to validate the various correction procedures used to obtain accurate material properties.

Extensive creep data of the type shown in Figure 1.26 have proved particularly valuable for modeling the influence of physical aging on creep behavior and for the prediction of long-term creep from short-term tests [18]. It is worth noting that, in contrast to the situation with amorphous polymers (Section 1.4), physical aging in semicrystalline materials at certain temperatures appears to produce reductions in relaxation magnitudes as well as shifts in relaxation time spectra [72]. The magnitude changes are similar to those produced by decreasing temperature (Section 4.4).

Acknowledgment

The authors wish to acknowledge discussions with Dr. A. F. Johnson relating to the theory in Section 3.1.

References

1. C. Zener, *Elasticity and Anelasticity of Metals*, Chicago University Press, Chicago, IL, 1948.
2. G. Bradfield, "Use in Industry of Elasticity Measurements in Metals With the Help of Mechanical Vibrations," *Notes on Applied Science*, No. 30, National Physical Laboratory, Her Majesty's Stationery Office, London, England, 1964.
3. J. D. Ferry, *Viscoelastic Properties of Polymers*, 3rd ed., Wiley, New York, 1980.
4. L. E. Nielsen, *Mechanical Properties of Polymers and Composites*, Dekker, New York, 1974.
5. N. G. McCrum, B. E. Read, and G. Williams, *Anelastic and Dielectric Effects in Polymeric Solids*, Wiley, London, 1967.
6. I. M. Ward, *Mechanical Properties of Solid Polymers*, 2nd ed., Wiley, Chichester, 1983.
7. B. E. Read and G. D. Dean, *The Determination of Dynamic Properties of Polymers and Composites*, Hilger, Bristol, 1978.
8. J. Heijboer, *Int. J. Polym. Mater.*, **6**, 11 (1977).
9. B. E. Read and J. C. Duncan, *Polym. Test.*, **2**, 135 (1981).
10. J. M. Dealy, *Rheometers for Molten Plastics*, Van Nostrand Reinhold, New York, 1982.
11. H. Markovitz, P. M. Yavorsky, R. C. Harper, L. J. Zapas, and T. W. DeWitt, *Rev. Sci. Instrum.*, **23**, 430 (1952).
12. W. P. Mason, *Trans. ASME*, **69**, 359 (1947).
13. A. J. Barlow, G. Harrison, J. Richter, H. Seguin, and J. Lamb, *Lab. Pract.*, **10**, 786 (1961).
14. J. F. Johnson, J. R. Martin, and R. S. Porter, "Determination of Viscosity," in A. Weissberger and B. W. Rossiter, Eds., *Techniques of Chemistry, Physical Methods of Chemistry*, Vol. 1, Part VI, Wiley, New York, 1977.
15. A. J. Staverman and F. Schwarzl, "Linear Deformation Behaviour of High Polymers," in H. A. Stuart, Ed., *Die Physik der Hochpolymeren*, Vol. 4, Springer-Verlag, Berlin, 1956.
16. B. E. Read, *Polymer*, **22**, 1580 (1981).
17. F. R. Schwarzl and L. C. E. Struik, *Adv. Mol. Relaxation Processes*, **1**, 201 (1968).
18. G. D. Dean, B. E. Read, and G. D. Small, *Plast. Rubber Process Appl.*, **9**, 173 (1988).
19. J. C. Snowdon, *Vibration and Shock in Damped Mechanical Systems*, Wiley, New York, 1968.
20. E. T. Clothier, *Plast. Rubber Mater. Appl.*, **1**, 41 (1976).
21. R. E. Wetton, *Appl. Acoust.*, **11**, 77 (1978).
22. R. Lane, M. R. Bowditch, and B. C. Ochiltree, *Prog. Rubber Plast. Technol.*, **1**, 61 (1985).
23. J. Lamb, "Thermal Relaxation in Liquids," in W. P. Mason, Ed., *Physical Acoustics*, Vol. II, Part A, Academic, New York, 1965.
24. B. E. Read, "Glass Transitions in Multicomponent Systems," in M. Pineri and A. Eisenberg, Eds., *Structure and Properties of Ionomers*, NATO ASI Series, C 198, Reidel, Dordrecht, 1987.

25. L. C. E. Struik, *Physical Aging in Amorphous Polymers and Other Materials*, Elsevier, Amsterdam, 1978.
26. B. E. Read, "Dynamic Mechanical and Creep Studies of PMMA in the α- and β-Relaxation Regions. Physical Ageing Effects and Non-linear Behaviour," in Th. Dorfmüller and G. Williams, Eds., *Molecular Dynamics and Relaxation Phenomena in Glasses, Lecture Notes in Physics 277*, Springer-Verlag, Berlin, 1987.
27. L. C. E. Struik, *Polymer*, **28**, 57 (1987).
28. R. H. Boyd, *Polymer*, **26**, 1123 (1985).
29. R. E. Wetton, *Polym. Test.*, **4**, 117 (1984).
30. G. D. Dean, J. C. Duncan, and A. F. Johnson, *Polym. Test.*, **4**, 225 (1984).
31. S. Etienne, J. Y. Cavaille, J. Perez, R. Point, and M. Salvia, *Rev. Sci. Instrum.*, **53**, 1261 (1982).
32. J. E. McKinney, S. Edelman, and R. S. Marvin, *J. Appl. Phys.*, **27**, 425 (1956).
33. A. F. Yee and M. T. Takemori, *J. Polym. Sci. Polym. Phys. Ed.*, **20**, 205 (1982).
34. A. R. Ramos, F. S. Bates, and R. E. Cohen, *J. Polym. Sci. Polym. Phys. Ed.*, **16**, 753 (1978).
35. D. J. Massa, *J. Appl. Phys.*, **44**, 2595 (1973).
36. T. L. Smith and W. Oppermann, *Tech. Papers 39th Annu. Tech. Conf. Soc. Plast. Eng.*, **27**, 163 (1981).
37. R. Hoshino, K. Kitamura, K. Murayama, and M. Todoki, Toray Research Center, Shiga, Japan, private communication.
38. K. Tanaka, Government Industrial Research Institute, Osaka, Japan, private communication.
39. D. J. Plazek, M. N. Vrancken, and J. W. Berge, *Trans. Soc. Rheol.*, **2**, 39 (1958).
40. A. E. Schwaneke and R. W. Nash, *Rev. Sci. Instrum.*, **40**, 1450 (1969).
41. ISO 537, *Plastics—Testing with the Torsion Pendulum*, International Organization for Standardization, Geneva, 1980.
42. J. Heijboer, *Polym. Eng. Sci.*, **19**, 664 (1979).
43. J. K. Gillham, *Polym. Eng. Sci.*, **19**, 676 (1979).
44. P. Sakellariou, E. F. T. White, and R. C. Rowe, *Br. Polym. J.*, **19**, 73 (1987).
45. L. C. E. Struik, *Rheol. Acta*, **6**, 119 (1967).
46. S. Yano, *J. Appl. Polym. Sci.*, **19**, 1087 (1975).
47. C. J. Nederveen and C. W. van der Wal, *Rheol. Acta*, **6**, 316 (1967).
48. G. D. Sims, G. D. Dean, B. E. Read, and B. C. Western, *J. Mater. Sci.*, **12**, 2329 (1977).
49. D. Bruce and P. Hancock, *J. Inst. Met.*, **97**, 140 (1969).
50. G. S. Radley and P. J. Banks, *J. Phys. E*, **14**, 546 (1981).
51. D. W. Robinson, *J. Sci. Instrum.*, **32**, 2 (1955).
52. F. J. Guild and R. D. Adams, *J. Phys. E*, **14**, 355 (1981).
53. C. J. Nederveen and F. R. Schwarzl, *Br. J. Appl. Phys.*, **15**, 323 (1964).
54. S. P. Timoshenko, D. H. Young, and W. Weaver, *Vibration Problems in Engineering*, 4th ed., Wiley, New York, 1974.
55. N. G. McCrum, *Polymer*, **25**, 299 (1984).

56. A. M. North, R. A. Pethrick, and D. W. Phillips, *Macromolecules*, **10**, 992 (1977).
57. R. A. Pethrick, *J. Phys. E*, **5**, 571 (1972).
58. T. Wright and D. D. Campbell, *J. Phys. E*, **10**, 1241 (1977).
59. A. J. Barlow, G. Harrison, and J. Lamb, *Proc. R. Soc. London Ser. A*, **282**, 228 (1964).
60. I. A. Viktorov, *Rayleigh and Lamb Waves*, Plenum, New York, 1967.
61. A. D. Cordellos, R. O. Bell, and S. B. Brummer, *Mater. Eval.*, **27**, 85 (1969).
62. T. L. Mansfield, *Mater. Eval.*, **33**, 96 (1975).
63. B. G. Martin, *Non-Dest. Test. (Guildford, Engl.)*, **9**, 199 (1974).
64. G. D. Dean, *Plast. Rubber: Process. Appl.*, **7**, 67 (1987).
65. M. J. P. Musgrave, *Crystal Acoustics*, Holden-Day, San Francisco, CA, 1970.
66. G. D. Dean and F. J. Lockett, *ASTM Special Technical Publication*, **521**, 326 (1973).
67. B. Hartmann and J. Jarzynski, *J. Acoust. Soc. Am.*, **56**, 1469, (1974).
68. J. E. Zimmer and J. R. Cost, *J. Acoust. Soc. Am.*, **47**, 795 (1970).
69. G. L. Petersen, B. Chick, and W. Junker, *IEEE Ultrasonics Symp. Proc.*, 650 (1975).
70. W. Sachse and Y. Pao, *J. Appl. Phys.*, **49**, 4320 (1978).
71. R. A. Kline, *J. Acoust. Soc. Am.*, **76**, 498 (1984).
72. B. E. Read, G. D. Dean, and P. E. Tomlins, *Polymer*, **29**, 2159 (1988).

Chapter **2**

DETERMINATION OF YIELDING, CRAZING, AND FRACTURE

Hugh R. Brown

This chapter is restricted to the consideration of the deformation and failure processes in polymeric materials. This restriction is made because a chemist is more likely to be concerned with measuring such properties in polymers than in metals or ceramics. There are, however, many similarities among the experimental techniques used for these three classes of materials, particularly in the area of fracture toughness testing. For more information about tests on nonpolymeric materials the reader is referred to standard textbooks [1, 2].

1   YIELD

1.1   Introduction

The concept of yield was originally developed for metallic materials to describe the situation where a sufficiently high stress causes nonrecoverable deformation. Anyone who has bent a steel wire such as a coat hanger is familiar with the concept of yield and plastic deformation. The deformation of the wire under small stress is elastic and can be recovered when the stress is removed, but the plastic deformation that occurs when the stress is high enough to cause yield does not recover when the stress is removed. It is not obvious that this concept of yield is as applicable to polymers as it is to metals. In general, in a metal, the stress–strain curve is fairly straight and independent of strain rate up to a finite strain. At this finite strain the metal begins to yield by the process of dislocation movement so that the strain is no longer fully recoverable when the load is removed. In some metals the stress continues to rise during the yielding process, but in other metallic materials load maxima and slip or yield bands are seen. In all cases the deformation is never recoverable. Yield stresses are often defined by an offset criterion; a line is drawn on the stress–strain curve parallel to the initial elastic line but through a point corresponding to an offset strain such as 0.2% on the abscissa. The yield stress is then defined as the stress where the stress–strain curve crosses this offset line.

In polymers the form of the stress–strain curve is fairly dependent on the testing rate but often resembles that shown in Figure 2.1. It is convenient to consider the low strain deformation in terms of viscoelastic theories and the deformation at and beyond the load maximum as yield. The yield stress can then be defined as the stress at the load maximum. This definition is clearly arbitrary; however, as it is convenient, it is commonly used when considering yield in polymers. It is clear from the viscoelastic, time-dependent nature of the deformation that definitions of yield based on offset strains or deviations of the

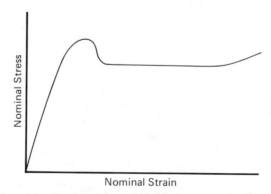

Figure 2.1   A typical stress–strain curve of a ductile polymer.

stress–strain curve from linearity are not as useful in polymers as they are in metals.

A material that shows a stress–strain curve of the form shown in Figure 2.1 deforms beyond the yield point by forming a neck, a localized region of high deformation. Further deformation occurs as the shoulders of the neck travel along the gauge length of the specimen transforming the undeformed material to highly oriented material. This process of cold drawing is found in many polymers, and the deformation ratio of the drawn material is called the *natural draw ratio*.

1.2   Yield Criteria

Yield criteria are equations that attempt to describe the yield point of a material in any stress state, and so can give, for example, the relation between the yield stresses in tension and compression. A yield criterion is a function that contains the different components of the stress tensor that describe the yield point. There are just two basic yield criteria that are commonly used in polymers, the *Coulomb yield criterion*, with its pressure-independent version named after Tresca, and the *Von Mises yield criterion*.

The Coulomb yield criterion states that yield will occur when the shear stress τ resolved onto any plane reaches a critical value that varies linearly with the normal stress σ on that plane. The criterion can be written as

$$\tau = \tau_c - \mu\sigma_n \tag{1}$$

where τ_c and μ are constants.

This relation is based on a frictionlike concept so that the shear stress required for yielding increases with the compressive stress across the plane.

The Von Mises yield criterion is best expressed in terms of the principal components of the stress tensor $(\sigma_1, \sigma_2, \sigma_3)$. It states that yield will occur when

$$(\sigma_1 - \sigma_2)^2 + (\sigma_2 - \sigma_3)^2 + (\sigma_3 - \sigma_1)^2 = 6K^2 \tag{2}$$

where K can be pressure dependent. Further discussion of yield criteria is found in [3].

It is necessary to measure the yield points in a series of different stress states to establish the yield criterion for a material. The stress states that can be obtained without too much difficulty are uniaxial tension, uniaxial compression, torsional shear, and biaxial tension. We can obtain uniaxial tension and compression directly by pulling or pushing rods or bars using the tensile testing machines that are discussed in more detail in Section 1.4. Torsional shear can be obtained by twisting either solid rods or thin-walled cylinders. Biaxial tension can be obtained by the inflation of a circular membrane. Some typical sample geometries are shown in Figure 2.2. The aim of the test is always to obtain the stress–strain curve of the material in the relevant stress state and hence find the

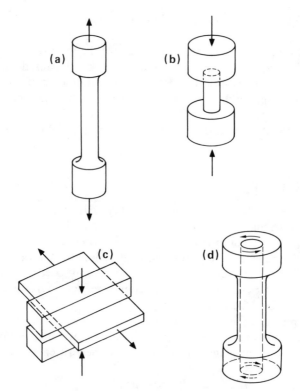

Figure 2.2 Sample geometries that are used for the study of yield criteria: (a) simple tension, (b) simple compression, (c) plane-strain compression, and (d) torsion.

stress at which yielding starts. In multiaxial or compressive stress situations it is often not possible to define the yield stress as the maximum stress because load drops normally do not occur. The stress–strain curves often show an obvious bend at the end of the linear elastic region, and the stress at which this bend occurs is often defined as the *yield stress*.

Another test that has proved convenient for the measurement of yield in a range of stress states is the plane strain compression test. In this test the sample is compressed between long narrow dies and simultaneously pulled normal to the long axis of the dies (Figure 2.2). The deformation is in the plane normal to the long axis of the dies. This geometry has been used to show that the yield of poly(methyl methacrylate) (PMMA) fits the Coulomb yield criterion [4]. A sample geometry that allows a range of stress states is the thin-walled cylinder where torsion can be combined with both tension and internal pressure. The disadvantage of this geometry is that both the testing machine and the samples become complex and expensive.

Relatively simple geometries have been used to examine the effects of hydrostatic pressure on yield. Normally, cylindrical samples have been tested in either tension or torsion while they are mounted in a pressure vessel and surrounded by a fluid under high pressure. The pressure was found to change the deformation from ductile to brittle or vice versa depending on the stress state and whether the sample is protected from the pressure fluid. It should be noted that protection of the sample from the pressure fluid changes the mechanics of the brittle failure. This is discussed in detail in [5].

1.3   The Effects of Temperature and Strain Rate on Yield

Polymer yield stresses normally increase with increasing strain rate and decreasing temperature. Most materials show a minimum temperature, for any given strain rate, for which the material will yield. At temperatures below this minimum, the material fails in a brittle manner at strains below the typical yield strain of about 5%. While this brittle-ductile transition temperature is the minimum temperature for yield, the maximum temperature is either the glass transition temperature T_g for amorphous polymers or the crystal melting temperature for semicrystalline polymers.

The effect of temperature and strain rate on yield has mainly been studied by using simple tensile elongation experiments. The results have often been found to fit an Eyring type equation for an activated rate process. This gives an equation for the strain rate $\dot{e}$

$$\dot{e} = \dot{e}_o \exp\left(-\frac{\Delta H}{RT}\right) \sinh\left(\frac{V\sigma}{RT}\right) \tag{3}$$

where ΔH is the activation energy, V is the activation volume, R is the gas constant, T is the absolute temperature, and σ is the yield stress. The sinh term can normally be approximated as an exponential giving an equation

$$\frac{\sigma}{T} = \frac{R}{V}\left[\frac{\Delta H}{RT} + \ln\left(\frac{\dot{e}}{\dot{e}_o}\right)\right] \tag{4}$$

This equation suggests that a plot of σ/T versus log (strain rate) will give a family of parallel straight lines from which the activation energy and activation volume can be obtained. An example, taken from [6], is given in Figure 2.3 for polycarbonate (PC). This model has been found to fit several materials over a limited temperature range, but it is not clear that the parameters obtained, particularly the activation volumes, help much in the understanding of the yield process.

This simple Eyring model can of course be generalized to allow for more than one process with the processes occurring in series or parallel. It can also be generalized by allowing the activation volume to be temperature dependent. Both approaches allow the fitting of data, but they do not seem to add much in the way of understanding the yield process.

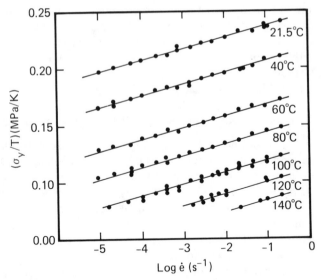

Figure 2.3 Yield data for polycarbonate plotted as yield stress over temperature versus strain rate to show the fit to (4). Reprinted with permission from C. Bauwens-Crowet, J. A. Bauwens, and G. Homès, *J. Polym. Sci. A2*, **7**, 735 (1969).

1.4   Yield Testing

Most yield tests are performed on standard commercial mechanical testing machines. These machines are designed to stretch a sample steadily and provide a record of the variation of the load on the sample with its deformation. Some of them can also apply loads or deformations to the sample that vary in a complex way with time. There are two basic classes of such machines—those that operate using servohydraulics and those that are based on lead screws. The latter type, the screw machines, are more common as they are cheaper and simpler to use. They operate by a constant speed motor turning two lead screws that are mounted in a stiff load frame. A crosshead that is mounted between the two screws is thereby moved up or down at constant speed. Typically, a load cell is bolted on the crosshead and then the sample is clamped between the crosshead and the load frame. The sample is hence stretched or compressed as the crosshead moves. The machines are normally switchable over a large range of crosshead speeds and load cell ranges. However, neither the load range nor the testing rate range available is as great for screw machines as it is for servohydraulic machines.

Servohydraulic machines do not have a movable crosshead in the same way. Instead a hydraulic actuator (basically a hydraulic ram) is mounted on the lower horizontal member of a stiff rectangular load frame. The sample is mounted between the ram and a load cell that is bolted to the upper horizontal member of the load frame. In this way ram movement is used to stretch or compress the sample. The ram is driven from a large hydraulic pump and its motion is controlled by a servovalve. The control system, which is normally referred to as closed loop, utilizes a feedback system so that the sample load, the ram motion, or the sample strain can be made to follow a programmed command. The servocontroller obtains the difference between the signal that represents the load, ram motion, or strain and the programmed input signal; amplifies it; and sends it as a control signal to the servovalve. In this way the difference signal is minimized so the sample load or deformation follows the program signal accurately. Any electrical signal can in principle be used as the program signal, but in practice it is normally obtained from a digital function generator or computer. Clearly, the parameter being controlled (load, ram movement, or sample strain) must be a single-valued function of the amount of hydraulic fluid in the actuator as the latter is what the servovalve really controls. Otherwise the feedback could go from negative to positive causing loss of control and, at minimum, destruction of the specimen. Constant rate of deformation, constant rate of loading, and fatigue tests are straightforward with such machines. However, if yielding or failure is being studied, operation in load or sample strain control can be problematic as the load drops both at the yield point and at the start of crack propagation. In these cases load and some local measures of sample strain can be multivalued functions of ram movement. Commerical servohydraulic machines are available that permit combined axial tension and torsional loading of the sample. Such machines are very useful for the examination of yield criteria. Clearly servohydraulic machines are much more sophisticated (expensive) and complicated to use than screw machines, but they do offer greater freedom to the experimenter.

Modern tensile testing machines are normally interfaced in some way to a computer. In the simplest systems the computer is just used as a data logger with convenient programs to do simple analysis of the results. In more sophisticated systems the machine is entirely controlled by the computer. Computer control is clearly worthwhile with some servohydraulic machines as the computer can be used to generate the control waveform. It is also valuable when many standard tests are being done as a test profile can decrease the chances of error. It is not obvious to me that computer control offers a great deal to the use of a screw-type machine in a research environment.

Studies of yield, crazing, and failure often require that mechanical tests are done at temperatures other than room temperature. Temperature control is normally accomplished by the use of purpose built ovens (often called "environmental temperature chambers") that mount on the load frame of the machine. They normally cool or heat the sample and offer temperature ranges from about -100 to $+200°C$ or sometimes up to $+350°C$. The temperature uniformity offered is typically $\pm 1°C$ with an accuracy of about $\pm 2°C$.

Mechanical testing equipment is offered by a number of suppliers, but the name most commonly associated with such equipment is the Instron Corporation.

1.5   Molecular or Structural Models of Yield

For a polymer to yield polymer chain segments must move past each other; therefore, the yield processes are likely to be strongly influenced by the local packing of polymer chains. The local packing of polymer chains is very different in amorphous polymers from that in the crystallites of semicrystalline polymers. The yield processes are most likely very different in these two classes of polymers. This is particularly the case when the amorphous regions in the semicrystalline polymers are in a rubbery state above their T_g and so produce little resistance to shear deformation, so that the process of yield basically occurs in the crystallites. We will first consider two models that have been proposed to explain the yield in glassy polymers and then go on to consider yield in semicrystalline polymers.

The Robertson yield theory [7] is based on the idea that the effect of stress on the polymer can change the ratio of bond conformations in the same way as occurs with an increase in temperature. An effective temperature is calculated and used in a Williams, Landel, and Ferry (WLF) equation to calculate a flow rate. Within this model the main resistance to deformation is intramolecular in origin. Recent infrared (IR) spectroscopic evidence shows that the ratio of trans to gauche bonds in poly(styrene) does change at yield in the manner suggested by the Robertson model [8]. It should be emphasized that there is no suggestion of a significant rise in actual temperature; the various adiabatic heating models of yield have been thoroughly disproved [3].

The Argon model of yield [9] is based on the concept that, at the molecular level, the yielding process occurs by the formation of pairs of molecular kinks. This process is shown in Figure 2.4. The resistance to kink formation is considered to come from the necessity to cause elastic strain in the surrounding material. The kink energy is calculated from the theory of elastic disclinations and used as the activation enthalpy required for a deformation, which is calculated from the kink dimensions. This model fits experimental data on yield stresses over a wide range of temperature and strain rates. The temperature and rate dependence of the low strain properties is already included in the modulus value that is used to calculate the strain energy of the kinks, and this helps the model fit the experimental data.

These two models of yield emphasize different aspects of the yield process. The first is concerned with intramolecular processes, while the second considers only intermolecular forces. There is most likely some truth in both models, with the Argon theory probably being most applicable at low temperatures and the Robertson process being more significant close to T_g.

The yield of semicrystalline polymers is a much more complicated problem than that of glassy polymers. The crystals are often organized in the form of spherulites, and there is every reason to suspect that the different regions of a spherulite will deform differently because a spherulite is a spherically symmetric aggregate of many crystallites. The crystallites grow out from the center of the spherulite with one particular crystal axis in the radial direction. Hence, the

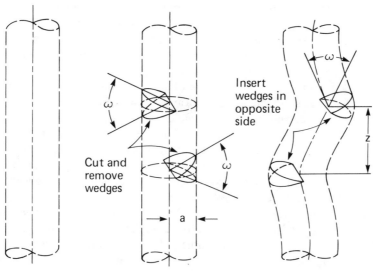

Figure 2.4 Modeling of the pair of molecular kinks as the formation of a pair of disclinations in Argon's model of yield in glassy polymers. The disclinations have an angle ω in a cylinder of radius a and are a distance z apart.

orientation of the crystallites with respect to the stress direction depends on their positions within the spherulite. However, the deformation processes of a crystallite are most likely to depend on the direction of the stress with respect to the crystal axes. Therefore, one would expect different deformation processes to occur in different regions of a spherulite. For example, if the stress is uniaxial tension, this can be represented by a vector and the deformation processes that occur in the poles of the spherulite (with respect to this vector) are different from those that occur at the equator. This problem has been examined by the application of a range of morphological techniques.

We can examine deformation on the scale of the spherulite by using optical microscopy and small-angle light scattering [10]. Deformation within the spherulite has been studied by electron microscopy and small-angle X-ray scattering (SAXS), but neither of the latter two techniques is particularly powerful in this application. Electron microscopy can only be done on thin films and can cause such rapid radiation damage that in situ deformation studies will be meaningless, while SAXS can only look at deformation averaged over several spherulites (unless the material contains unusually large spherulites) and so can give no detailed local information.

In view of this lack of experimental information it is not surprising that many deformation processes have been proposed. These range from models that are heavily based on the deformation processes that are known to be significant in metallic systems, such as dislocation motion, twinning, and martensitic transformations, to models that assume that the polymer crystallites deform by partial melting. Dislocations have been shown to exist in polymer single crystals and are probably important in the deformation of such samples. Twinning has also been seen; however, the importance of these processes in the deformation of bulk, melt crystallized material is unclear, although a model based on screw dislocation motion does seem to work fairly well in poly(ethylene) [11]. There is good evidence that, in some regions of the spherulites, interlamellar slip is important; whereas, in the equatorial and polar regions (with respect to the stress direction) considerable intralamellar reorganization and voiding must occur. Partial melting has been detected in at least two ways: (1) the lamellar spacing of a yielded and drawn material has been found to depend entirely on the drawing temperature and not on the lamellar spacing of the undrawn starting material, and (2) the degree of segregation of deuterated polyethylene in a matrix of hydrogenated poly(ethylene) has been found to change on yielding and drawing the sample [12, 13]. It is hard to imagine how this can happen without some partial melting. The most complete picture of the processes of yielding and drawing of semicrystalline polymers is that proposed by Peterlin [14]. This model has the problem of being based on concepts of highly regular chain-folded lamellae that we now know do not exist in melt-crystallized material.

In conclusion, it is clear that the yielding and drawing processes of semicrystalline polymers are complicated and not well understood.

2   CRAZING

2.1   Introduction

Crazing is a localized mode of deformation that occurs only in amorphous polymers below T_g or in semicrystalline polymers below T_m. Crazes look like cracks to the naked eye; in fact, they are load-bearing structures. Their internal structure consists of fibrils of highly oriented polymer that are oriented across the craze with the interfibrillar region being either void, or, in the case of environmentally induced crazes, active fluid. An electron microscope image of a craze in poly(styrene) is shown in Figure 2.5. Typical fibril diameters are 5–25 nm and volume fraction of the fibrils in the craze is in the range from 0.2 to 0.5. It is difficult to learn about the mechanical properties of the fibrils because they are so small. However, indirect evidence shows that their tensile properties are influenced by their large surface-to-volume ratio; so the fibrils are more elastic than bulk highly oriented polymer [15]. Crazes normally initiate at internal or surface flaws and inhomogeneities and grow normal to the maximum principal stress [16]. There are many craze growth criteria in the literature but none fits the experimental data as well as yield criteria fit yield data. The craze criterion proposed by Oxborough and Bowden [16] is probably the most useful.

A second type of crazes called *crazes II* or *intrinsic crazes* occurs in some polymers, particularly PC, when the polymer is highly oriented and close to the glass transition [17]. These crazes, which have very large fibrils, are probably associated with entanglement loss by chain slippage.

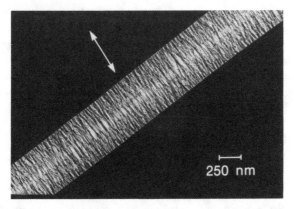

Figure 2.5 A typical transmission electron micrograph of a craze in poly(styrene). The arrow shows the stress direction.

2.2   The Crazing Processes

Crazes occur in polymers but not in other materials because of the way the long-chain nature of polymers affects their stress–strain properties. Consider polymers as consisting of entangled chains with the degree of entanglement sufficiently severe that the entanglement points cannot slip greatly at temperatures below T_g. As shown in Figure 2.1, polymers in tensile deformation tend to strain soften beyond the yield point; then, at high deformation, as the sections of the chains between entanglements begin to straighten, the load rises again and the material strain hardens. The strain softening tends to localize plastic deformation, and the strain hardening suppresses ductile failure. The localized regions of plastic deformation are constrained from elongating in one direction and contracting in the other two directions by the less deformed regions around them, and they release the constraint by forming voids.

The most successful model of the craze growth process is the meniscus instability model that was proposed by Argon (see Argon and Salama [18]) and has been developed by Kramer [19]. In this model the polymer is considered a highly viscous (perhaps plastic) fluid that is constrained to flow in a narrow gap. The thin layer of fluid is constrained by the elastic material around it, so the process is rather akin to the pealing of Scotch tape off glass. The model predicts the fibrillar type of morphology observed and suggests that the mean fibril diameter $\bar{D}$ should relate to surface energy Γ and crazing stress S in the form

$$\bar{D} = 8 \, \frac{\Gamma v_f^{0.5}}{S} \tag{5}$$

where v_f is the fibril volume fraction. The surface energy contains a chain breakage term that can be very significant. This term is also important in environmental crazing.

The strength of a craze and the craze's stress are both strongly affected by the extension ratio of the crazed matter, which is related to the craze fibril volume fraction by volume conservation. A craze with highly oriented fibrils will tend to have low v_f and so be weak. The extension of craze fibrils has been shown to relate to the maximum possible extension of chains between entanglements as does the extension of cold drawn material. The extension ratio of crazes, however, tends to be a little larger than that of yield zones [19].

2.3   Environmental Effects in Crazing

Crazes frequently occur in glassy polymers in the presence of organic fluid environments. The organic fluid is believed to work mainly as a plasticizer reducing the local flow stress of the polymer, but the reduction in surface energy also has some effect. For some polymers a good correlation exists between reduction in crazing stress and solubility parameter of the fluid. Fluids with solubility parameters close to that of the polymer tend to dissolve the polymer and are cracking agents. As the solubility parameter difference increases, the crazing strain increases. For polar materials one must consider the different components of the solubility parameter.

Semicrystalline polymers also craze in the presence of some swelling agents. The most studied situation is the environmental stress cracking of poly(ethylene) in soaps and detergents. This can be a serious problem in practice because it is unexpected. It probably occurs by a stress-induced swelling and plasticization process. The poly(ethylenes) most resistant to this form of failure (and probably most brittle failure) are the high molecular weight medium density materials of which gas pipe grades are an example. It would appear that a high density of interlamellar tie molecules is a requirement for cracking resistance.

The most aggressive crack-inducing materials are chain-breaking agents. A couple of classic examples for PC are potassium hydroxide dissolved in ethanol and aqueous solutions of ethanolamine. Both solutions produce brittle cracks with weak rudimentary crazes. Many other examples exist but they tend to be specific to the particular polymer chemical structure. One that has caused several problems is the cracking of nylons in divalent metal salts. The crazing of acrylics in ultraviolet light is another example. The cracking of elastomers in ozone also occurs by a chain-breaking process.

2.4   Techniques for Studying Crazes

2.4.1   Mechanical Techniques

The main aim of mechanical measurements related to crazing has been to find the stress or strain criteria for craze initiation and growth. Simple tensile tests have been used; but as craze initation is a time-dependent phenomenon, they are not very suitable. Constant load creep experiments are more useful as they can give time to craze as a function of applied stress. As crazing criteria are multiaxial, it is necessary to use sophisticated specimen shapes and loading systems to examine them. Oxborough and Bowden [16] obtained a known biaxial stress pattern by using a sample containing a cylindrical hole. The sample was loaded in tension normal to the cylinder axis and in compression along the cylinder axis. To find more complete multiaxial information Duckett and co-workers [20] have developed a rig that can load a hollow cylinder in axial tension (or compression) together with twist and internal pressure. The mechanical tests used to examine craze criteria can be performed on a tension–torsion servohydraulic machine of the type discussed in Section 1.4.

The effect of hydrostatic pressure on crazing has been studied by loading solid cylinders in tension or torsion together with the hydrostatic pressure. The equipment used is specialized and rather complicated as the generation and sealing of high pressures is not easy. The reader is referred to the original literature for experimental details [20, 21]. The interpretation of the tension data has proved problematic because the most reasonable fluids used as the pressurizing medium can be environmentally active; but if the sample is protected from the fluid, the craze cannot fill with fluid and the stress state becomes unclear [5].

Probably the most useful mechanical data obtained on crazes are on the effects of an external environment on crazing. The most convenient test involves clamping the sample to an elliptical jig so that the maximum strain varies in a well-known way along the sample and then immersing the sample and jig in the relevant fluid. The sample is then left for 24 h and the minimum strain for crazing found by direct observation of where on the sample the crazes stop [22].

2.4.2 Optical Techniques

The area of optical examination of crazes includes both optical microscopy, which is useful and straightforward, and craze refractive index determination. From a measurement of craze refractive index Kambour [23] first showed that crazes, which were known at the time to be load bearing, were highly voided structures. He measured the refractive index difference between the craze and the uncrazed polymer by finding the critical angle for total internal reflection. This technique is very simple because only a protractor and a light source are required. It is useful when one has a few large crazes in macroscopic samples, and it is a good way to find if, for example, environmental (solvent) crazes are filled with fluid. The density of polymer within the craze v_f is normally found from the craze refractive index if we assume the Lorentz–Lorenz relation and the additivity of the polarizabilities of the fibrils and the air or solvent around them. For samples that are a few micrometers thick we can find the craze refractive index by using transmitted light interference techniques with an optical microscope. Jenoptik Jena makes an optical microscope togther with an interferometer that is very convenient for measuring the refractive index of small samples.

Reflection optical microscopy has proved very useful in the study of crazes at crack tips [24]. The craze and crack are observed from above (normal to the crack plane) so that interference bands are produced by the light reflected from the upper and lower craze surfaces in a manner similar to the formation of Newton's rings. A typical example is given in Figure 2.6; this photograph was taken by using a standard microscope with no special equipment although long working distance objectives can be useful here. These patterns can be used to obtain stress and deformation information about the crack tip craze and hence to learn more about the micromechanics of crazing and crack propagation. The interference technique has proved particularly useful in PMMA, but it has been employed in most of the rigid glassy polymers [25].

2.4.3   Electron Microscopy Techniques

Transmission electron microscopy (TEM) is a very valuable technique for the study of crazes and the crazing process. The resolution is excellent and normally controlled by the specimen preparation rather than by basic instrumental limitation. Hence, any reasonably modern TEM is suitable for the examination of crazes. The main disadvantage of the technique is that thin samples are required. For unstained polymers film thickness's of about 400 nm are about optimum although thickness's of up to $1 \mu m$ can be used in high voltage microscopes. These thicknesses are perhaps, at first sight, surprisingly large; but it should be remembered that crazes typically have a v_f from 0.25 to 0.5 and so contain much less matter than the bulk polymer. Other difficulties of this

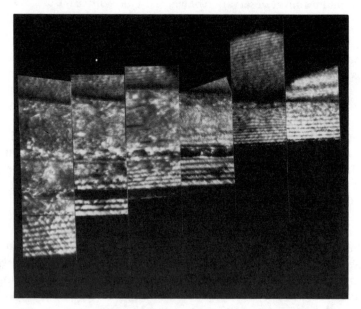

Figure 2.6 Reflection optical micrographs of the crack tip craze formed in fatigue in poly(vinyl chloride). The crack and craze are growing from top to bottom and a series of six pictures of the same area taken at successive times are shown. The crack tip craze is made evident by the series of clear interference bands within it.

technique are related to the considerable radiation damage that occurs in the microscope. The damage makes experiments in some unstable materials, such as PMMA, very difficult and also makes in situ experiments difficult to interpret.

Two different methods are commonly used to make the thin samples that are suitable for TEM. In the first method a bulk sample is crazed, and then thin slices are microtomed from the crazed block. Often the crazed sample is stained with osmium tetroxide and thereby cross-linked between the crazing and the microtomy [26]. The staining and microtomy are fairly standard in the microscopy of polymers, and the techniques are described in some detail in reviews of polymer microscopy [27]. The second method consists of forming a thin film from solution, attaching it to a copper grid and then straining the copper grid, thereby crazing the film [28]. Thin films can be made from a solution either by spinning using a standard resist spinner by drawing a glass slide at a constant rate from a beaker of the solution. The films are floated off the substrate and picked up on a copper grid that has been coated with a thin layer of the polymer. The grid carrying the film is then placed in the solvent vapor for a short time so that the film bonds to the grid and tightens. The sample is then dried in a vacuum oven at a temperature below T_g and then stretched in a testing machine or a little strain frame based on a motor driven micrometer positioner. Both methods of sample preparation have their uses. The second method can only be used in polymers that do craze when in the form of thin films, and it is not obvious that the structure of the crazes formed in thin films will be typical of crazes in bulk materials. The microtomy method permits the study of crazes formed in bulk materials; but as crazes are relatively weak structures, there is a good chance that they will be severely deformed by the microtomy process. Staining decreases this problem but adds the possibility that some of the crazes were formed during the staining process.

The main uses of TEM have been to examine the fibrillar morphology of crazes; to study craze intiation and breakdown, particularly in rubber toughened polymers; and to find local values of the fibril volume fraction v_f and craze dimensions. The parameter v_f is important because it gives the extension ratio of the craze fibrils and, with the craze dimensions, can be used to calculate craze stresses. The interpretation of the craze images in terms of the morphology has problems that can be partially overcome by using the microscope to form low-angle electron diffraction patterns [29].

Scanning electron microscopy (SEM) has also been useful in the examination of crazing; but since it is a surface tool of lower resolution than TEM, its use has been more limited. It is used mainly to study crazes on fracture surfaces, but often the resolution has not been good enough to see the craze fibrils. This should not be the case with modern microscopes that have resolutions of 3 nm or better.

2.4.4 Small-Angle Scattering Studies

Crazes are in some ways ideal structures for study by SAXS techniques as there is a large contrast between the fibrils and the void and the size scales are about optimum. Both slit and point collimated radiation have been used; but because the scattering pattern from crazes is highly anisotropic, the interpretation of either type of pattern requires care. One problem that has caused confusion in the literature arises because the scattering patterns consist of both diffraction from the fibrils and reflection from the craze–matrix interfaces. The analysis of these two components has been discussed in detail in a recent review [30].

The simplest SAXS technique for the study of crazes is to use a slit collimated radiation, which is perhaps most easily obtained from a Kratky camera. It is necessary to keep the sample at the crazing strain during the experiment so there must be room for a load frame around the sample to hold it in either tension or three-point bend. An old style Kratky camera with its separate collimator block and flight tube is convenient in this regard. The design of SAXS cameras and methods of data analysis are discussed in detail in a recent book [31].

Small-angle scattering has been used mostly to study fibril sizes in crazes in bulk polymers. When either X rays or neutrons are used, the optimum sample thickness is about 1 mm. High intensity X rays available from a synchrotron source have made real-time studies possible and both fatigue and impact processes have been examined [30]. In most cases the mean fibril size has been obtained using Porod analysis, which involves measuring both the small-angle scattering invariant (a measure of the total amount of scattering) and the variation of scattering intensity with angle at high angles. A knowledge of the primary beam intensity together with the invariant can be used to measure the volume fraction of crazes in the volume examined by the beam. Small-angle scattering using electrons, as mentioned in the Section 2.4.3, has been used to measure the mean fibril diameter of TEM samples at thicknesses where the images are hard to interpret and hence compare the results from such thin samples with those obtained from bulk materials [29].

3 FRACTURE

3.1 Introduction

The fracture properties of polymers can be measured in many ways depending both on the nature of the materials and the form of fracture that is of interest. For example measurement of impact failure in a brittle plastic is very different from measurement of static fatigue (creep fracture) in a rubber. The most fundamental distinction, however, is not between different materials or different rates but between tests on sharply notched samples and tests on unnotched samples. The tests on notched samples can normally be interpreted in terms of fracture mechanics, but only in exceptional circumstances can fracture mechanics be used to analyze unnotched tests.

3.2 Unnotched Tensile Tests

In general the primary fracture data that are used to characterize any polymer are obtained from a tensile test. The normal information that can be obtained on almost any material includes stress and strain to break. The nature of the stress–strain curves can, however, cause certain difficulties in the interpretation of these figures. If the failure strains are below about 3%, the stress–strain curve is linear, the material is brittle, and the failure tends to be flaw controlled. This means that specimen preparation, specimen shape, and even specimen size can affect the results. Parallel-sided strips often break in the clamps giving erroneous results, but samples with wider clamping regions can often break at the stress concentration regions where the sample narrows. The surface conditions of the sample can be very important with very smooth samples giving significantly higher failure strains than slightly rough samples. The effect of sample volume or surface area on failure strains is well recognized in the testing of inorganic glasses where statistical (Wiebull) analysis is often used. The basic principle of this approach, which has also been applied to glassy polymers, is that a distribution of flaw sizes exists so the larger the sample, the greater the chance of a large flaw.

Interpretation of stress and strain to break can also be a problem in ductile polymers. Such materials normally form a neck that then propagates along the sample at constant load. The deformation is therefore highly inhomogeneous, and a strain obtained from the crosshead movement of a tensile testing machine is invalid. In addition, the question of whether a neck will propagate all along the gauge length of a sample is often controlled by flaws. The process of initial neck formation is often rather unstable; the sample load decreases and so the nonnecked portions of the sample undergo elastic retraction. Deformation that was lost from these elastic regions concentrates into the region forming a neck. This can mean that the local strain rate that occurs on neck formation can be unrelated to the crosshead speed and controlled by the length of the sample; that is, the length of elastically strained materal that is being partially unloaded. As samples can often fail during neck formation it is clear that such failure is more likely in long samples, where this effect is more important, than in short samples. When a neck forms, it will most commonly propagate all along the gauge length so relatively small changes in sample dimensions can change apparent failure strains from values typical of neck formation, 10–15%, to the natural draw ratio of the polymer, 150–300%, with little change in failure stress.

Some tough polymers continue to elongate after a neck has enveloped all of the sample gauge length; high molecular weight medium and low density poly(ethylene) grades are good examples of materials that do this. Such high strain deformation is normally homogeneous and so meaningful failure stresses and strains can be obtained.

These problems that are connected with inhomogeneous deformation are not an issue for elastomers. Meaningful and reproducible results are normally obtained on rubbers when the considerable problems of clamping have been solved. It is interesting to note that Smith [32], in his extensive work on the effects of strain rate and temperature on failure of rubbers, normally tested rings that did not have to be clamped. Some of the tough high-temperature polymers, of which the polyimide pyromellitic dianhydride-4,4'-oxydianiline (PMDA-ODA) (Kapton H) is a good example, tend to deform homogeneously without necking and so give reproducible and meaningful failure strains. The results obtained are still easily influenced by surface flaws.

From the preceding discussion it should be evident that failure stresses and strains obtained from tensile tests must be interpreted with caution. The whole stress–strain curve gives much more information on the failure properties of the material than just a failure stress and strain. It also gives an idea of the amount of nonelastic deformation and the strain hardening that the material can sustain before fracture.

We normally conduct tensile testing on the machines discussed in Section 1.4. The problems most commonly found in practice are connected with sample clamping, and for this reason a range of different clamp styles and clamp faces are available. Clamp faces range from rubber to sandpaper to roughened steel; in addition, it is possible to make samples that have wedge or conical shaped ends so that they lock positively into specially shaped clamps.

Several standard sample shapes are defined in the American Society for Testing and Materials (ASTM) standards, and it is often advisable to use one of them. As mentioned earlier, the condition of the sample surface can be very important in controlling the results obtained. Surface contamination, as well as roughness, can have a large effect on the results obtained. For example, if the polymer is susceptible to environmental crazing, then oils obtained from manual handling can cause premature crazing and failure, particularly in low-speed, high-temperature tests.

The ASTM and other national standards organizations have specified some standard tensile tests, and to produce trade literature it is necessary to follow such prescriptions [33].

3.3   Fracture Mechanics

It is evident from the previous discussion that the failure of materials is often controlled by flaws. This is even more the situation in real parts than in tensile test situations as real parts often contain stress concentrations, molding flaws, or minor surface damage. Clearly this flaw-controlled failure cannot be defined by concepts such as a macroscopic failure stress and so another approach is required. Fracture mechanics has been developed as the other approach and is concerned with the characterization and understanding of the propagation of preexisting flaws or cracks.

There are two basic approaches to fracture mechanics. The first approach supposes that a crack will propagate when there is sufficient strain energy available to create the two crack surfaces. The energy required is not just the surface energy of the material but also includes a much larger term that describes the dissipative processes that occur around the crack tip. This total energy required is described by G_c, the critical strain energy release rate, that is defined as an energy per unit area of crack growth. In principle, G_c can be calculated for any loaded test piece or part and so can be used as a failure criterion.

The other approach to fracture mechanics is based on the realization that the singular elastic stress field close to a sharp crack tip is not influenced by the shape of the body and can be described entirely by a parameter K, called the stress intensity factor. It is reasonable then to assume that the failure; that is, the propagation of this sharp crack, will occur when K reaches a critical value K_c. The fracture toughness K_c is hence assumed to be a material property. These two approaches appear distinct but can be shown to be equivalent in situations where the material remote from the crack tip is linearly elastic. When the material shows large-scale nonlinearities (elastic or plastic) the energy approach has normally been found to be more useful.

The basic aim of fracture mechanics is to characterize the toughness of materials in terms of G_c or K_c, which hopefully are independent of specimen dimensions and sample type. These toughness values can then be used both to compare materials and to calculate the tendency for crack propagation to occur in practical sample geometries. The area of fracture mechanics of polymers recently has been reviewed in detail [34, 35].

An issue that is important in any approach to the testing of notched samples is the form of the stress and deformation patterns round the crack tip. These patterns are essentially three-dimensional but are normally considered in one of two important limits—plane stress and plane strain. Plane stress is the situation where the only significant stresses are those within the plane of the sample; the through thickness stresses are assumed to be zero. In plane strain it is the through thickness deformations and strains that are assumed to be zero, and from Poisson's ratio effects, there are significant through thickness stresses. The deformation at a crack tip changes from plane stress to plane strain as the sample thickness increases. This concept, and particularly the existence of plane strain deformation, becomes particularly important when considering plastic deformation at the crack tip. It is evident from the form of yield criteria, particularly (2), that the maximum principal stress to cause yield will increase with the through thickness stress. Hence, plastic zones are smaller and toughnesses are lower in plane strain than in plane stress. In addition, in many polymers the mode of deformation at the crack tip can change from yielding to crazing as the sample thickness increases. This issue can be very important when considering notch sensitivity of materials and the relation between test results and field failures.

3.3.1   The Energy Balance Approach

The energy balance approach to fracture mechanics is based on the concept that a crack will propagate when the total energy available per increment of crack area G is greater than or equal to that required for the new crack area. If the

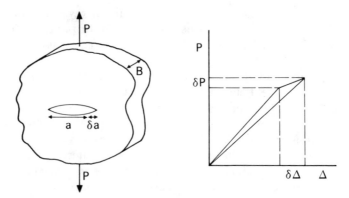

Figure 2.7   Schematic diagram of a cracked sample. The crack of length a grown by δa in a sample of width B experiencing a load P. The crack growth causes the deformation Δ to grow by $\delta\Delta$ while the load increased by δP.

available strain energy in the sample is U and the work done by the loading system is W, then

$$\frac{1}{B}\frac{d(W-U)}{da} \equiv G \geqslant G_{\mathrm{c}} \tag{6}$$

where the crack has length a and the sample has a width B. For purely brittle materials the crack growth resistance G_{c} is equal to 2γ, where γ is the surface energy of the material. For most materials, including all polymers, G_{c} is very much greater than γ.

LINEAR ELASTIC MATERIALS
When the bulk of the sample is linearly elastic, then, from Figure 2.7, one can consider U_1 as the strain energy before an increment of crack growth and U_2 as the value after the crack has grown by an increment δa. The external load is P and the deflection of the loading points is Δ. Then

$$U_1 = P\Delta/2 \tag{7}$$

and

$$U_2 = (P + \delta P)(\Delta + \delta\Delta)/2 \tag{8}$$

The work performed by the loading system during this increment is

$$\delta W = (P + \delta P/2)\delta\Delta \tag{9}$$

Therefore, the total energy available is

$$\delta(W - U) = (P\delta\Delta - \Delta\delta P)/2 \tag{10}$$

So

$$G = \frac{1}{2B}\left(P\frac{\delta\Delta}{\delta a} - \Delta\frac{\delta P}{\delta a}\right) \tag{11}$$

The crack will propagate when $G \geqslant G_c$ so G is a feature of the geometry and loads and its critical value G_c is hopefully a material constant. It is valuable to put (11) in terms of the specimen compliance C

$$C = \Delta/P \tag{12}$$

so that

$$\delta\Delta = P\delta C + C\delta P \tag{13}$$

and hence

$$G = \frac{P^2}{2B}\frac{dC}{da} \tag{14}$$

Equation (14) has proved very useful as the compliance of the cracked sample cannot only be measured experimentally, but it can also be calculated using linear elasticity for many sample geometries.

NONLINEAR ELASTIC AND PLASTIC MATERIALS

The energy balance approach to fracture mechanics is not restricted to linear elastic materials. It has proved to be a powerful technique for both nonlinear elastic (rubbery) polymers and materials that show large-scale plasticity. In fact the earliest applications of fracture mechanics to polymers were made by Rivlin and Thomas (see [35]) in their classic studies on crack growth in cross-linked rubbers. The approach customarily used for rubbers is very similar to that given previously for linearly elastic materials, but the phraseology is slightly different. For rubbers it is assumed that the crack growth increment occurs with fixed loading grips (constant Δ) so that the crack will grow when

$$-\frac{1}{B}\left(\frac{\partial U}{\partial a}\right)_\Delta \geqslant G_c \equiv T \tag{15}$$

Here T is called the tearing energy of the rubber and is equivalent to G_c. It is assumed that a is the length of the crack in the relaxed sample. As $U(a, \Delta)$

$$dU = \left(\frac{\partial U}{\partial a}\right)_\Delta da + \left(\frac{\partial U}{\partial \Delta}\right)_a d\Delta \tag{16}$$

Hence,

$$\left(\frac{\delta U}{\partial a}\right)_\Delta = \frac{dU}{da} - \left(\frac{\partial U}{\partial \Delta}\right)_a \frac{d\Delta}{da} \tag{17}$$

This result is useful for the calculation of G for specific samples. For example, it is straightforward to calculate G for the trouser leg or simple extension sample shown in Figure 2.8. If the crack grows a length da at a constant applied force P, then a length da of material is converted from unloaded to loaded. The strain energy density in the uniformly loaded arms is W_0, and the arms have an original (unloaded) cross-sectional area A_0. Hence,

$$dU = W_0 A_0 da \tag{18}$$

by virtual work

$$P = \left(\frac{\partial U}{\partial \Delta}\right)_a \tag{19}$$

If the extension ratio $(1 + e)$ of the legs is λ, then

$$d\Delta = 2\lambda da \tag{20}$$

$$G_c = \frac{2\lambda P - W_0 A_0}{B} \tag{21}$$

The sample is normally designed so that the deformation in the legs is insignificant so

$$G = \frac{2P}{B} \tag{22}$$

There is a large body of work on the application of the energy approach to fracture mechanics to other inelastic situations. The most used parameter is the J integral developed by Rice. The J integral approach becomes equivalent to the approach using G as the nonelastic region of the sample decreases. It is based on a formula identical to that just discussed for rubbers, but its application is more involved and specific to particular materials and test types. It will not be considered more here, and for additional information the reader is referred to more specialized treatments [34].

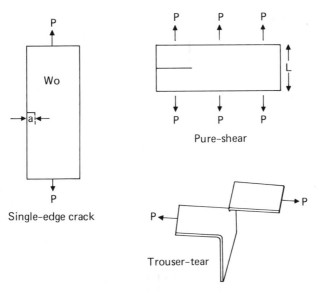

Figure 2.8 The geometry of some samples that are used in the fracture mechanics of elastomers. The samples experience a load P and W_0 is the strain energy density in the sample.

3.3.2   The Stress Intensity Approach

The stress intensity factor approach to fracture mechanics is based on an analysis of the stresses that exist around the tip of a sharp crack in a loaded elastic solid. These are normally described using two-dimensional polar coordinates (r, θ) whose origin is at the crack tip so the original crack is at $\theta = \pi$ (see Figure 2.9). The crack can be loaded in three modes, known as I, II, and III, which correspond to crack opening and in plane and antiplane shear, respectively. Mode I is by far the most important and is the only mode that will be considered here. The stresses around the crack tip are given by

$$\sigma_{ij} = \frac{K_{\mathrm{I}}}{(2\pi r)^{1/2}} \, f_{ij}(\theta) \qquad (23)$$

where σ_{ij} are the components of the stress tensor and f_{ij} represents a function of θ. It is evident from this equation that the stress situation close to the crack can be described entirely by the stress intensity factor K_{I}. The sample geometry and loading can only alter the stress through a change in value of K_{I}. Also, the stresses approach infinity at the crack tip so it is clear that stress itself cannot be a reasonable fracture criterion. It is therefore reasonable to assume what the condition for crack propagation will be when K_{I} is greater than or equal to a critical value K_{Ic}. The opening mode is often assumed; hence, this is just given as K_{c}.

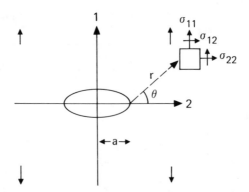

Figure 2.9   The geometry of stresses around crack tips. The applied tensile stress is in the 1 direction normal to the crack of length $2a$. The stresses σ are those experienced in a volume element a distance r from the crack tip at an angle θ to the 2 direction.

Both the K_c and G_c approaches to fracture mechanics are based on linear elasticity and it is not hard to show that they are equivalent. If opening mode is again assumed, then

$$G_c = \frac{K_c^2}{E^*} \qquad (24)$$

where E^*, the *reduced modulus*, is equal to E for plane stress and $E/(1 - v^2)$ for plane strain.

3.3.3   Fracture Mechanics Testing

Fracture mechanics tests are conceptually simple to perform. However, considerable care is often necessary to obtain results of high accuracy and precision. The basic process consists of selecting a sample shape, initiating a crack at the correct point within the sample, mounting the sample in the testing machine, and loading it at a constant rate until the crack propagates in an unstable manner. We can then obtain K_{Ic} or G_c from the load at unstable crack propagation together with the calibration for the particular sample geometry. In most nonpolymeric materials unstable crack propagation is normally observed with perhaps just small amounts of stable (slow speed) crack propagation at increasing K_I before the crack becomes unstable and travels rapidly across the sample. In polymers long-distance stable crack propagation at constant K_I can often be found, so fracture mechanics testing is often used to find the relationship between K_I (or G) and crack velocity. Critical values of K_I or G can still often be defined as the crack normally becomes unstable at velocities of perhaps a few millimeters per minute. Poly(methyl methacrylate) is a very convenient example of a material where the crack growth is stable over a wide velocity range but unstable propagation still occurs at high crack speeds. Its

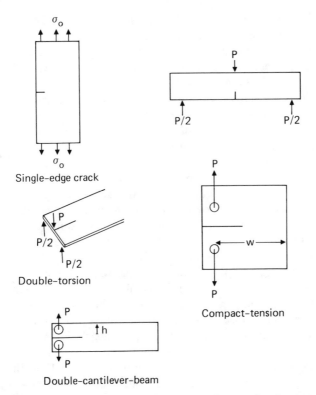

σ_0

Single-edge crack

Double-torsion

Double-cantilever-beam

P

P/2   P/2

P

w

Compact-tension

P

h

Figure 2.10 Geometries of some typical fracture mechanics samples that are used for rigid polymers. For the single-edge crack sample σ_0 is the remote stress. The other samples experience a load P.

fracture properties have been well studied [34] so it is useful for checking one's testing techniques. Unstable crack propagation is not normally seen in either the testing of elastomers or the environmental stress cracking of rigid polymers, so in both cases the aim of a fracture mechanics test is to obtain the relation between K_1 and crack velocity.

A range of sample geometries are commonly used for fracture mechanics testing of polymers. A few examples are given here, but the reader is referred to more specialized textbooks for more information [2, 34, 35]. For testing of rubbers the most common geometries are probably the simple tension (trouser-tear) test piece already discussed and the single-edge notch sample. For the latter sample G is given by

$$G = 2kW_0a \qquad (25)$$

where $k \simeq \pi/\lambda^{1/2}$. In this relation W_0 is again defined as the strain energy density in the sample remote from the crack. For small deformations ($\lambda \simeq 1$)

(25) tends to the infinitesimal elastic result for a crack in an infinite sheet. For more rigid materials the most common geometries are given in Figure 2.10. The calibration factors of these samples can be found in the literature [35]. The type of sample used depends on the nature of the material being tested. For example fracture mechanics has proved to be very useful in the analysis of environmental stress cracking in poly(ethylenes), but these materials are by no means linearly elastic. However, as long as the tests are restricted to samples that do not have a long lever arm, consistent results can be obtained. Single-edge, double-edge, and center-notch samples have proved most useful in this situation [36]. A mechanically stable specimen geometry is required to characterize slow steady crack growth situations. The compact tension and double torsion samples are most useful in this regard. The compact tension sample has the extra advantage that it does not need to be grooved; the crack tends to travel along the center line. However, the double torsion sample, which does require grooving, has the advantage that the relation for K_c or G_c does not contain the crack length, so the crack travels at constant applied load. In relatively ductile materials it is necessary to use thick samples to constrain the size of the plastic zone at the crack tip (this will be discussed in more detail later), and the three-point bend and compact tension samples are most useful in this regard.

When we use fracture mechanics, we assume that a preexisting sharp crack is present in the sample. For some materials, particularly PMMA, this does not cause any problem as it is easy to form a sharp (single craze) crack just by fatigue or by carefully pushing a clean razor blade into the sample. For other materials if a crack is originally set up with a bundle of crazes at its tip, it will propagate with the bundle of crazes and give much higher values of K_{Ic} than if it had been carefully set up with just a single craze at the tip. Poly(styrene) will act this way. Whether this is considered just an experimental problem that requires care to avoid or evidence that K_{Ic} is not a material constant is a matter of opinion.

It is found experimentally that many materials give K_{Ic} or G_c values that depend on the sample thickness. The values decrease as the sample thickness is increased until a constant value is obtained for thick samples. This variation comes about because the plastic (or craze) zone at the crack tip is larger at the surface of the sample than it is in the center. The plastic zone size is influenced by the through thickness stresses, which are zero at the sample surface, and build up in the interior of the sample to a constant value in the region where the deformation is plane strain. In most polymers the plane strain plastic zone is in fact a craze or a bundle of crazes, but yielded plastic zones can be seen on some polymer sample surfaces. Valid fracture toughness values are normally considered to be those obtained with plane strain deformation at the crack tip. There is a criterion for valid fracture toughness tests borrowed from the metals literature that is often used for polymers

$$B \geqslant 2.5 \left(\frac{K_{Ic}}{\sigma_y} \right)^2 \tag{26}$$

where σ_y is the yield stress and B is the sample thickness. With materials that form a single crack tip craze valid results can be obtained with sample thickness considerably less than those given by this relation.

The variation of observed K_{Ic} or G_c with sample thickness can sometimes show an interesting instability. This instability, normally known as a brittle ductile transition, has been most studied in PC in impact tests but is evident in other materials It is often observed as a sudden increase (by a factor of 5 or more) in impact strength as the sample width is decreased or, alternatively, as the testing temperature is increased. This phenomenon has been explained as a coupling of the surface plastic zone size with the mean fracture toughness of the sample where the toughness is influenced by the plastic zone size [37].

3.4   Impact Testing

Impact tests are among the most commonly used fracture tests on rigid polymer materials. The tests fall into two broad categories: sharp (or relatively sharp) notch tests and unnotched or blunt indenter tests. Falling dart tests are an example in the latter category; these tests are not considered further here as there is no clear way of obtaining fundamental material parameters from them. The sharp notch tests are normally done by mounting the sample and then hitting it with a pendulum striker. The fundamental measurement is the energy lost by the pendulum in breaking the sample, but frequently the pendulum is also instrumented so that load-time curves can be obtained. There are two basic test geometries. Charpy tests are essentially four-point bend tests, but the two center loading points are relatively close together, so the tests are similar to notched three-point bend tests. In an Izod test the notched sample is mounted so that it can be considered as a built-in cantilever and then its free end is hit by the pendulum. Standards that define sample dimensions include notch-tip radius and sometimes it is obviously necessary to adhere to such standards. However, more information on the failure of a material or on comparisons between materials can normally be obtained by studying the effect of sample dimensions and notch-tip radius on impact failure.

Sharp-notched impact tests can be considered as high-speed fracture mechanics tests and used to obtain G_c values. This is normally done by a simple extension from (14). The elastic energy in the sample at failure is given by

$$U = G_c BD\phi \tag{27}$$

where D is the sample width and

$$\phi = C/[dC/d(A/D)] \tag{28}$$

so the calibration constant can be calculated easily for standard Izod or Charpy samples. They are to be found in [34]. This approach has assumed that the kinetic energy lost by the pendulum goes into the strain energy of the sample at failure, but in reality the situation is more complicated. Some energy goes into the kinetic energy of the sample, and complex multiple impact effects have also been analyzed in some detail [34]. The use of instrumented machines has removed the necessity of the energy analysis and permitted the examination of the shapes of the loading curves, but it has its own set of problems because serious ringing effects are often present in the transducers. In principle the best approach is to instrument the sample; however, for obvious reasons, this is not often done.

References

1. A. H. Cottrell, *The Mechanical Properties of Matter*, Wiley, New York, 1964.
2. J. F. Knott, *Fundamentals of Fracture Mechanics*, Butterworth, London, 1976.
3. I. M. Ward, *Mechanical Properties of Solid Polymers*, 2nd ed., Wiley, Chichester, 1983.
4. P. B. Bowden and J. A. Jukes, *J. Mater. Sci.*, **3**, 183 (1968).
5. R. A. Duckett, *J. Mater. Sci.*, **15**, 71 (1980).
6. C. Bauwens-Crowet, J. A. Bauwens, and G. Homès, *J. Polym. Sci. A2*, **7**, 735 (1969).
7. R. E. Robertson, *J. Chem. Phys.*, **44**, 3950 (1966).
8. M. Theodorou, B. Jasse, and L. Monnerie, *J. Polym. Sci. Polym. Phys. Ed.*, **23**, 445 (1985).
9. A. S. Argon, *Philos. Mag.*, **28**, 839 (1973).
10. J. M. Schultz, *Polymer Materials Science*, Prentice-Hall, Englewood Cliffs, NJ, 1974.
11. R. J. Young, *Mater. Forum*, **11**, 210 (1988).
12. W. Wu and G. D. Wignall, *Polymer*, **26**, 661 (1985).
13. G. D. Wignall and W. Wu, *Polym. Commun.*, **24**, 354 (1983).
14. A. Peterlin, *J. Polym. Sci.*, **69**, 61 (1965).
15. H. R. Brown and E. J. Kramer, *Polymer*, **22**, 687 (1981).
16. R. J. Oxborough and P. B. Bowden, *Philos. Mag.*, **28**, 547 (1973).
17. M. Dettenmaier, *Adv. Polym. Sci.*, **52/3**, 57 (1983).
18. A. S. Argon and M. Salama, *Mater. Sci. Eng.*, **23**, 219 (1976).
19. E. J. Kramer, *Adv. Polym. Sci.*, **52/3**, 1 (1983).
20. R. A. Duckett, B. C. Goswami, L. S. A. Smith, I. M. Ward, and A. M. Zihlif, *Br. Polym. J.*, **10**, 11 (1978).
21. K. Matsushige, E. Baer, and S. V. Radcliffe, *J. Macromol. Sci. Phys.*, **B11**, 565 (1975).
22. R. P. Kambour, E. E. Romagosa, and C. L. Gruner, *Macromolecules*, **5**, 335 (1972).
23. R. P. Kambour, *J. Polym. Sci. Macromol. Rev.*, **7**, 1 (1973).
24   H. R. Brown and I. M. Ward, *Polymer*, **14**, 469 (1973).
25. W. Döll, *Adv. Polym. Sci.*, **52/3**, 103 (1983).
26. O. S. Gebizlioglu, R. E. Cohen, and A. S. Argon, *Makromol. Chem.*, **187**, 431 (1986).
27. S. Y. Hobbs, *J. Macromol. Sci. Rev. Macromol. Chem.*, **C19**, 221 (1980).

28. B. D. Lauterwasser and E. J. Kramer, *Philos. Mag. A*, **39**, 469 (1979).

29. H. R. Brown, *J. Polym. Sci. Polym. Phys. Ed.*, **21**, 483 (1983).

30. H. R. Brown, *Mater. Sci. Rep.*, **2** 315 (1987).

31. O. Glatter and O. Kratky, *Small Angle X-ray Scattering*, Academic, London, 1982.

32. T. L. Smith, *Rubber Chem. Technol.*, **51**, 225 (1978).

33. *Annual Book of ASTM Standards*, American Society for Testing and Materials, Philadelphia, PA, 1988.

34. J. G. Williams, *Fracture Mechanics of Polymers*, Wiley, New York, 1984.

35. A. J. Kinloch and R. J. Young, *Fracture Behaviour of Polymers*, Applied Science, London, 1983.

36. K. Tonyali and H. R. Brown, *J. Mater. Sci.*, **21**, 3116 (1986).

37. H. R. Brown, *J. Mater. Sci.*, **17**, 469 (1982).

Chapter **3**

THE MEASUREMENT OF FRICTION AND WEAR

Norman S. Eiss Jr.

1   INTRODUCTION

Humans have shown an interest in controlling and measuring friction and wear since recorded history. Ancients recorded in their paintings [1] the number of animals or humans required to pull a load across the ground, and thus they provided a crude measure of friction force. Paintings in pyramids record how the ancients reduced friction by pouring a liquid on the ground in front of a sliding platform that was used to carry a statue. The invention of the wheel was motivated by a desire to reduce the friction of sliding by replacing sliding with rolling. Chariot wheels show evidence that animal fats were used to reduce friction between the hub and the axle [2]. While ancients were not concerned about measuring wear, a person of those times most likely noticed the ever-deepening grooves in the paving stones over which the chariots rode or the change in contour of stone steps over which millions of feet had trod.

Leonardo da Vinci performed the first recorded friction experiments in 1499 from which he formulated two empirical laws of friction: the friction force is proportional to the normal load and the friction force is independent of the apparent area of contact between the two bodies. His findings went unnoticed for 200 years until the industrial revolution stirred interest in understanding the cause of friction and subsequently controlling it. In 1699 Amonton confirmed da Vinci's findings, and he and his French contemporaries and successors postulated that the cause of friction was the roughness of surfaces in contact [2].

In 1734 Desauglier found that when two lead balls are pressed together and twisted they stick together. This adhesion was proposed as another cause for the force of friction. In 1804 Leslie suggested that friction is caused by the energy dissipated in the permanent deformation of materials when they are slid one on another. The study of friction in the eighteenth and nineteenth centuries concentrated on measuring friction of a variety of materials under dry and lubricated conditions [2].

The level of interest in measuring wear paralleled the increase in the level of accuracy in manufacturing machine components. In steam engines of the eighteenth century clearances between a piston and cylinder could be as large as 5 mm. Consequently, the loss of a few milligrams of material had no significant impact on the performance of the engine. However, when machine tools were capable of producing clearances between pistons and cylinders of 25 μm, the loss of a few milligrams of material produced a significant change in the machine performance [3]. Measurement of small amounts of wear also required advances in instrumentation. Radioactive tracers were used to measure small amounts of wear in gasoline engines without disassembling the engine [4]. Analytical balances capable of measuring tenths of micrograms and electronic gauging tools with sensitivities of hundredths of a micrometer made the measurement of microscopic mass losses and dimemsional changes possible. Scanning electron microscopes revealed the topographical changes resulting from wear, and a variety of analytical surface chemistry instruments detected chemical changes resulting from sliding.

As a result of these advances in measurement and analysis, the literature on friction and wear of materials grew exponentially with time. Unfortunately, our ability to predict friction and wear from basic principles is still in its infancy. Most of the literature consists of reports of the measurement of friction and wear, and occasionally the authors propose a model that predicts the changes observed in friction and wear as a result of changes in the variables of the experiment. Generally the models break down when we try to extrapolate beyond the test conditions of the experiment. Consequently, for each new application we must measure the friction and wear if these values are critical for the proper operation of the component.

Most books on tribology (the study of interacting surfaces in relative motion) include a discussion about the measurement of friction and wear. One of the most comprehensive treatments of this subject has 536 references [5]. Other informative treatments include those by Peterson [6] and Czichos [7]. The Society of Tribologists and Lubrication Engineers (formerly the American Society of Lubrication Engineers) published a compendium of friction and wear devices that are used worldwide [8].

This chapter provides information to help an experimenter plan an experiment to measure friction or wear and to interpret the results. A summary of the current mechanisms of friction and wear is followed by a discussion of experimental objectives. Methods of measuring friction and wear and the components of laboratory devices are then presented. The test procedures and the analysis and presentation of the results concludes this chapter.

2   THEORIES AND MECHANISMS

In all sliding contacts a normal load is transmitted across the contact. In most systems; for example, bearings, cams, and gears, the major function of the contact is the transmission of this normal load. Regardless of the initial topography of the surfaces, the normal load causes deformation of the surfaces to create an area of contact across which the load is transmitted. This *real area* of contact is normally a small portion of the *apparent area* of contact that is defined by the boundaries of the macroscopic interface of the two surfaces. As sliding occurs the materials on each member in contact have relative motion. Thus at one instance of time an element of one material can be in intimate contact with the other material, and in the next instance it can be in contact only with the gaseous or liquid environment. Hence, the area of contact is dynamic as new elements of the materials enter and leave the contact area. The dynamics of the area of contact and the interaction of chemical and mechanical phenomena complicate the understanding of the causes of the friction forces and the wear of the materials.

2.1   Friction

In the contact region bonding forces exist between the atoms of materials in close proximity. In all systems van der Waals type bonds can be found. If the materials are metals and the deformation of the contact has broken oxide layers or removed adsorbed gases, then metallic bonds can be formed [9]. Consequently, when sliding motion occurs the bonds between sites on either side of the contact are strained and either are broken or transmit forces to weaker bonds in either or both of the contacting materials. One component of the force required to slide one body over another is caused by the adhesive bonds formed in the contact area. This component is appropriately called the *adhesion component of friction*. A simplified model of the adhesion component of friction is that the friction force is the product of the area of contact and the shear strength of the weaker of the two materials in contact [10].

Because surfaces have roughness it is possible that the formation of the contact area involves the high points on one surface (*asperities*) that are geometrically interfering with those on the other surface. As relative motion occurs the interfering asperities deform and possibly fracture. The force required for this is called the *plowing component of friction*. For many years these two components of friction were considered to be independent, and in all but the softest of metals (e.g. lead and indium) the plowing component of friction was assumed to be insignificant compared to the adhesion component [10].

More recently Johnson [11] proposed that the two components are dependent. Qualitatively, it is argued that an increase in the adhesive bonds causes higher stresses to be transmitted across the contact area, which increases the deformation of the materials. The deformation can cause oxide films on the surfaces to rupture, which exposes metallic atoms capable of much stronger adhesive bonds.

When sliding occurs, the friction force moves through a distance; that is, the friction force does work. It is a common observation that sliding surfaces increase in temperature. Thus it is proposed that the frictional work is dissipated as heat in the materials adjacent to the contact area. The mechanisms of energy dissipation are plastic deformation, elastic hysteresis, and crack propagation [12].

In the preceding discussion of friction relative motion between the surfaces in contact was assumed. This friction is called *kinetic friction* to distinguish it from the force required to initiate the sliding, which is called *static friction*, and is higher than kinetic friction. One explanation for the higher value of static friction is that during static contact diffusion can occur at the real areas of contact and the number of adhesive bonds can increase. Consequently, the force required to break these bonds is greater than that required to break the bonds formed in dynamic contacts [13].

A second explanation is based on the observation that microslips occur during the loading that precedes sliding [14]. These microslips occur as the asperities in contact deform in response to the load. Each microslip proceeds until a set of asperities is encountered that can withstand the applied load. This process repeats until there are no further asperity sets that can withstand the load; then sliding occurs. During sliding the friction force varies as different asperity sets are encountered. However, most friction force measuring systems cannot follow these rapidly changing forces and thus they are averaged to a value (kinetic friction) that is less than the static friction force.

2.2   Wear

Wear is defined as the progressive loss of material from a surface occurring as a result of relative motion at the surface. Several mechanisms of wear are defined in Table 3.1. Very often the loss of material is preceded by the transfer of material from one surface to another. This transfer is accompanied by a change in surface topography that is caused by the loss or gain of material and by plastic deformation of the surfaces. The formation of a wear particle must be preceded by a fracture process.

The fracture process is controlled by the magnitude of the stresses necessary to propagate cracks in the presence of the surrounding environment. If the stresses caused by the normal and friction forces are high enough, a single application of them can cause existing cracks to propagate until fracture occurs, and a wear particle forms. This process is called *abrasive wear* because it is most commonly associated with wear caused by abrasive particles such as aluminum oxide or silicon carbide. These sharp-edged particles cause very high stresses

Table 3.1   Wear Mechanisms[a]

Mechanism	Definition
Abrasive wear	Wear by displacement of material caused by hard particles or protuberances
Adhesive wear	Wear by transference of material from one surface to another during relative motion due to a process of solid phase welding
Corrosive wear	A wear process in which a chemical or electrochemical reaction with the environment predominates
Fatigue wear	Removal of particles detached by fatigue arising from cyclic stress variations
Fretting	Wear phenomena occurring between two surfaces having oscillatory motion of small amplitude
Fretting corrosion	A form of fretting in which chemical reaction predominates

[a]Reprinted, with permission, from A. W. J. deGee and G. W. Rowe, "Glossary of Terms and Definitions in the Field of Friction, Wear, and Lubrication; Tribology," in M. B. Peterson and W. O. Winer, Eds., *Wear Control Handbook*, American Society of Mechanical Engineers, New York, 1980, pp. 1143–1201. [15].

when forced against other materials during sliding. Abrasive wear can occur whenever a rough surface is harder than the surface it contacts.

Stresses that are not high enough to cause abrasive wear can cause plastic deformation. The examination of an abrasively ground surface shows ample evidence of material that was plowed out of furrows. This material can be plastically deformed many times before it becomes detached as a wear particle. On a typical abrasive surface, only a small percentage of the abrasive particles are in a correct orientation and sharp enough to remove material by abrasive wear. The remaining particles are plowing grooves in the surface [16].

At an even lower level of stresses, the deformations in the materials are primarily elastic. Under these conditions elements of the surfaces passing in and out of areas of contact experience a reversal of elastic stresses that eventually leads to the formation and propagation of cracks. These cracks coalesce and cause a wear particle to be formed. Appropriately enough, this type of wear is called *fatigue wear*.

High stresses in the surface can also occur during sliding if the adhesive forces result from metallic bonds. Hence, the stresses can cause a fragment to be torn out of one surface and carried along by the other. This transfer wear can occur many times until a layer of material is created that is a mechanical mixture of the two materials in contact. It is from this mechanically mixed layer that the wear particles are eventually formed and ejected from the system. This transfer wear is usually called *adhesive wear* and it is most likely to occur when metals slide on metals [17].

A theory was proposed to explain the platelike wear particles observed in sliding experiments of metals on metals. The surface stresses cause subsurface cracks to form. The cracks are extended parallel to the surface during subsequent stress cycles. Ultimately, the crack extends to the surface and a platelike wear particle is detached from the surface. This theory is called the *delamination theory of wear* [18].

In some systems chemical reactions occur between the materials in the surface and the components of the environment, and the wear particles consist of the reaction products. This wear mechanism is called *corrosive wear*. A special case of corrosive wear called *oxidative wear* occurs when ferrous materials produce wear particles that consist of oxides of iron. In this oxidative wear mechanism it is postulated that adsorbed oxygen reacts with the iron to form iron oxide. As the oxide layer thickens, the iron and oxygen must diffuse through the layer and the rate of reaction reduces. At a critical thickness, the oxide layer fractures at the oxide–substrate interface, and a wear particle is formed that is largely composed of oxides of iron. The freshly exposed surface is then oxidized, and the process continues [19].

One form of wear is a complex combination of mechanical and chemical mechanisms. It is called *fretting corrosion*, and it occurs when the motion is in reciprocation with an amplitude that is small compared to the dimension of the contact area in the direction of the motion. The exact mechanism is unknown, but is consists of the exposure of reactive surfaces by fracture and the subsequent reaction of those surfaces with the environment and the abrasive action of the debris that is trapped in the interface [20].

In most sliding systems, all of the foregoing mechanisms occur. They may occur in some sequence or simultaneously. For example, a process called *running-in* is observed in most sliding systems. This refers to the observation that the wear rate at the beginning of the experiment is different from that measured later in the test. Often, this later wear rate is constant; hence, it is called *steady-state wear*. An example in which the wear rate during running-in is higher than the steady-state wear rate occurs when the running-in is caused primarily by the removal of the highest asperities of each surface. The steady-state wear then comprises some combination of the preceding mechanisms.

2.3   Magnitudes of Friction and Wear

Many systems show a linear relationship between friction force and normal load. The coefficient of friction is defined as the friction force divided by the normal load. For surfaces sliding in air without any intentional lubrication, coefficients of friction range from 0.05 to 1.5. This rather restricted range of values is explained by a model that predicts that frictional behavior is a function of the properties of adsorbed films rather than the properties of the underlying surfaces [21].

The lower values of the coefficient of friction are obtained for poly(tetrafluoroethylene) sliding on highly polished metal or glass surfaces. The higher values occur when soft metals like aluminum slide on each other [22]. The low friction results from the minimal plowing by the polished surface, the low adhesive forces between the molecular chains and the smooth surface, and the low shear strength of the polymer. The higher values occur when surface stresses cause the oxides to break up thus exposing aluminum atoms that can develop strong metallic bonds across the interface.

Unlike friction, the range of wear rates observed covers multiple orders of magnitude [23]. The lowest wear rates are observed for systems where the surface stresses are well below the yield strength of the material, chemical reactions between the materials and the components of the environments are minimal, and the adhesive bonds are of the van der Waals type. High wear rates occur during abrasive wear and sharp abrasives are always available in the contact area. High wear also occurs when strong metallic adhesive bonds can form.

3   EXPERIMENTAL OBJECTIVES

Friction and wear are measured for the following reasons: (1) the value of friction in a tribological system must be known to evaluate its performance or to size a power source, (2) the value of wear of a component is required to determine its useful life, (3) friction and wear measurements are made in studies of the physical and chemical processes that are responsible for friction and wear, and (4) measurements are also necessary when the tribological properties of different materials are to be compared. These measurements can be made in the field on systems where the tribological components are required for the proper function of the system or in laboratory devices.

3.1 Field Measurements

In systems that contain tribological components either or both of the values of friction and wear must be known to assess system performance and durability. In applications where the primary function is the transmission of a friction force, the value of friction is required to evaluate the component performance. For example, the friction torque in a brake must be known as a function of the brake design, materials, and environmental conditions. Likewise, the friction between a rubber tire and a road must be known to predict the performance of the vehicle. Another class of systems in which precise knowledge of friction forces is needed is in paper feeding systems. In many of these systems the friction of a pusher against the paper separates one sheet from a stack and moves it to a new position. In this application the friction force between the pusher and the paper must be greater than that between the adjacent sheets of paper so that only one sheet is removed each time. In addition, the friction force must remain constant over millions of operations and in a variety of temperature and humidity conditions.

In many other systems, such as cams and bearings, the primary functions of the systems are the transmission of a normal load between sliding members and the guidance or transmission of motion. However, the parasitic friction force contributes to the overall torque required to drive the system. Therefore, the friction torque must be known so that the power requirements for the system can be determined. This is particularly true for components used in outer space where overall weight must be minimized and available power is limited. For components such as slip rings, which are used to transmit electrical signals between parts that have relative motion, the friction torque in bearings and between the slip rings and brushes represents major fractions of the torques to be supplied by the drive motors.

Knowledge of the amount of wear in a tribological component is required to predict the life of the component. The life of a self-lubricated plain bearing is determined by the amount of radial wear that occurs before the shaft location has shifted to a point where the system can no longer perform its desired function. The increased clearance between the bearing and shaft could cause undesirable vibrations, gears to mesh improperly, or drive belts to reduce tension and slip on their pulleys. The life of a cam is determined by the amount of wear that changes the cam profile enough to cause the follower motion to deviate from specified tolerances. The wear of a cam in an automobile engine is a result of complex interactions between the materials of the cam and the follower, the additives in the lubricating oil, the duty cycle that affects the loads transmitted between the cam and follower and the sliding speeds, and the cam shape and valve spring that also affect the loads. These interactions change with operating time. Consequently, it is impossible to devise a laboratory test that will predict the wear of the cam in the engine. Thus, an engine test becomes the only method to give the wear life of the cam.

There are many problems associated with the in situ measurements. The tribological components are often inaccessible, which makes the measurements of friction difficult or impossible. Machines must be stopped and disassembled to make wear measurements. The wear rates are low so that tests require hundreds of hours to obtain measurable wear.

The measurement of friction and wear in the field is usually justified if the production volume of the components is large and the consequences of incorrect values of friction and wear in terms of expense, safety, and public relations are unacceptable. For example, field testing automobile engines to determine the wear of the critical components (cams, piston rings, cylinder walls, valves, and bearings) is standard practice in the industry. Testing in the field is also justified when the production volume is low but the system cost is extremely high. Equipment used in the exploration of outer space is extensively tested in vacuum chambers on earth to predict friction and wear characteristics accurately when the system is in space.

Because of the time and expense of field testing the wear of components, several methods are used to obtain the wear data in as short a time as possible. Wear can be measured with a high resolution system so that a small amount of wear measured in a short time can be extrapolated to predict the wear over much longer times. Of course, the actual wear rate must be constant over the extrapolated time if this method is to predict wear accurately. In the second method, either the normal load or sliding speed or both are increased so that more wear occurs in a given time interval. In this method the results will predict the wear in the field if the wear varies linearly with both load and speed. However, there are many systems in which the relationship among wear, load, and speed is nonlinear [24].

Accelerated tests do not correlate well with field tests for many reasons. Increases in sliding speed can increase frictional heating, and the mechanical properties of the component may change with temperature increase. The temperature increases may increase the rate of chemical reactions and thus change the mechanism of wear. Increases in load can increase stresses and cause the wear to change from fatigue to plowing or abrasive wear.

While field testing is often justified to verify that the friction and wear of tribological components meet expected performance and durability, it usually cannot be used to screen materials for the field because of the expense of testing. Materials are usually screened by a simple laboratory device that can produce measurable wear in short times at much less cost. Because field data are difficult to obtain there are very few examples where test results on laboratory devices have had a positive correlation with the measurements made in the field. Consequently, the field test remains the only reliable method that is available to obtain friction and wear values for the tribological components in the system.

3.2   Laboratory Devices

Primarily, laboratory devices are used to measure friction and wear (1) to study the basic processes that cause friction and wear, (2) to compare tribological performance of different materials, and (3) to predict the changes in friction and wear that will occur in an application as a result of changes in system parameters or operating conditions.

The mechanisms of friction and wear are studied largely on laboratory devices. The laboratory devices are designed so that a particular mechanism is a major factor in the wear. For example, abrasive wear is studied in devices where one of the sliding members is abrasive paper (this is called *two-body abrasion*). In another device abrasive particles are introduced between the test specimen and an elastomeric surface (*three-body abrasion*). Fatigue wear is predominant in a device where a hard sphere is loaded against a polymer surface and the nominal surface stresses are below the yield strength of the polymer [25].

Because the study of mechanisms requires as much information about the tribological system as possible, these devices are designed with the following features:

1. The normal load and sliding speed are accurately known and can be varied readily.
2. Specimens can be removed easily for measurement of dimensions or weight, visual observation, and chemical analysis.
3. Specimens are sized so that they can be placed in electron microscopes or surface chemical analysis instruments.
4. The environment (ambient temperature and gas composition) can be controlled and sometimes it can be analyzed directly.

Laboratory devices are used predominately to compare the tribological performance of materials. Laboratory devices used for evaluating different materials do not need to be as versatile as those used to study basic processes. The major requirement for the devices is that the conditions and test procudures must be the same from test to test. Standardized test machines and test procedures were developed by the American Society for Testing and Materials (ASTM) (see Section 9), the Society of Automotive Engineers (SAE), and other organizations. These tests give not only the details of the test machine, specimen preparation, operating conditions, and collection and analysis of the data, but also the coefficients of variation that can be expected. These data were developed through extensive round-robin interlaboratory tests prior to the issue of the standard. Very often standard test samples that can be used to calibrate a test apparatus are available from the National Institute of Standards and Technology (NIST).

The use of laboratory test devices to predict the tribological performance in an application is not possible because, by definition, the laboratory device is different in one or more significant aspects from the application. However, laboratory devices are used extensively as screening devices for materials for tribological applications. A screening test is used to identify materials that would have the best performance in an application at a cost much less than that required to test the materials in the application. Unfortunately, there are very few examples where laboratory test results correlate with applications. One such example is the positive correlation between the ranking of materials in a crossed-cylinder laboratory test with the ranking in a bearing test. The wear of carbon fiber reinforced polymer bearings caused by a steel shaft rotating for 500 h correlates with the wear of the outside of the bearing sliding against a steel cylinder in a crossed-cylinder configuration for 12 h. While the scatter was too large to predict the wear of the bearing from the crossed-cylinder test, the trend was used to select the materials with the lower wear rates [26].

Simulative testing [27] is a technique that has resulted in good correlations between laboratory tests and applications. The principle of simulative testing is that wear in the normal test can be studied by a laboratory test if the wear mechanisms are the same in both conditions. In this technique, the wear that occurs in a system component is characterized by measurement of the worn surface topography, microscopic observation of the worn surfaces to determine the surface morphology, and use of chemical analysis. If the same characteristics are observed in the laboratory test, the laboratory test can be used to study the effects of material changes, operating conditions, and environment on the wear of the component. This technique was used to devise a continuous sliding test that duplicates the failure of the magnetic coating on a rigid disk caused by the magnetic head during start-stop operations [28].

In the next sections the details of friction and wear measuring techniques are followed by a description of the essential features of laboratory devices.

4 TECHNIQUES OF MEASUREMENT

Because friction is the force required to initiate or sustain sliding, the principles of force measurement are applicable. Wear is usually measured by the mass lost from a component or calculated from a dimensional change caused by the loss of mass.

4.1 Measurement of Friction Force

One of the most common principles for measuring force is to measure the displacement of an elastic body that is deflected by the force. This principle is used almost exclusively to measure friction. There are many techniques used to measure the deflection of the elastic member.

In the friction measurement one of the sliding elements is mounted on an elastic member as shown in Figure 3.1. The deflection of the elastic member is measured by strain gauges, eddy currents, capacitance, interferometry, or piezoelectric crystals. The advantages and disadvantages of these techniques are detailed in books on instrumentation such as [29].

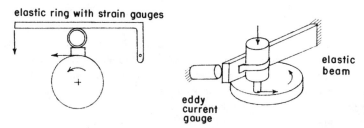

Figure 3.1 Techniques for measuring friction force.

In some cases one member is mounted on a pivoted arm. A spring is used to restrain the arm, and the deflection of the arm is measured. In this system, the friction torque at the pivot is a component of the measurement. The pivot friction force is minimized by using high-quality ball bearings that have very low rolling resistance.

The major disadvantage of measuring friction by the deflection of an elastic member is that the elastic member can store potential energy. Because the combined mass of the elastic member and the element attached to it has kinetic energy as a result of its velocity, the combination of the mass and elastic member is a system that will vibrate. There are two friction phenomena that result from this vibration, *stick-slip motion* and *self-excited oscillations* [30].

Stick-slip motion is characterized by alternating periods of sliding (slip) and nonsliding (stick) between the elements in contact. It usually occurs at low sliding velocities, and it is often associated with chatter and squeal emanating from the interface. Examination of the sliding surface reveals a pattern of repeating areas where stick and slip have occurred. Stick-slip can be prevented by increasing the sliding velocity, increasing damping in the elastic member, and increasing the stiffness of the elastic member. If a stiffer elastic member is used, then the deflection that is used to measure the friction force is reduced and a more sensitive displacement measuring technique is required.

While increasing the velocity eliminates stick-slip, it does not prevent the occurrence of self-excited vibrations. These vibrations can occur whenever the material in contact has a friction force that varies inversely with the relative sliding velocity. While the friction of metals is fairly insensitive to velocity, the friction of polymers does vary with velocity. A typical friction-velocity relationship shows the friction rising to a peak and then dropping as the velocity increases (see Figure 3.2).

The self-excited oscillations are initated by a disturbance that causes the element mounted on the elastic member to be displaced from its equilibrium sliding position. This motion causes the relative sliding velocity to change. For materials whose friction is independent of or increases with velocity, the elastically mounted element oscillates but the amplitude decreases and gradually approaches the equilibrium condition. The rate of approach is determined by the damping in the system. For the material with a negative slope to the friction–velocity curve, the system is unstable, and the oscillations grow and are

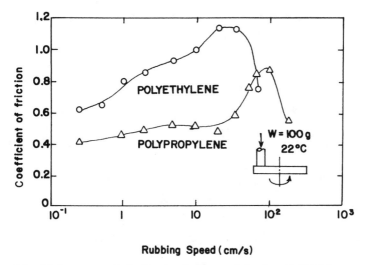

Figure 3.2   Friction versus velocity. Reprinted with permission from K. G. McLaren and D. Tabor, "Friction of Polymers at Engineering Speeds: Influence of Speed, Temperature, and Lubricants," *Proceedings of the Lubrication and Wear Convention 1963,* Institution of Mechanical Engineers, London, 1964, pp. 210–215.

sustained. To eliminate self-excited oscillations change the sliding conditions so that the material has a positive slope to the friction–velocity curve, increase the damping in the elastic support arm, and increase the stiffness of the support.

Because the elastically mounted element has a natural frequency of vibration, it also has a range of frequencies of excitation over which the friction force can be measured accurately. This range depends on the accuracy desired and the natural frequency, and it can be estimated from (1) for the response of the second-order system to a harmonic forcing function [32].

$$\frac{XK}{F} = \frac{1}{\{[1 - (\omega/\omega_n)^2]^2 + (2\zeta\omega/\omega_n)^2\}^{0.5}} \tag{1}$$

where X is the deflection of the member, K is the spring constant of the member, F is the friction force, ω is the frequency of the friction force, ω_n is the natural frequency of the element and the elastic member, and ζ is the damping ratio.

This equation is illustrated in Figure 3.3. Also shown is a band that indicates a desired accuracy, $XK = (1 \pm 0.05)F$. Note that for a zero damping ratio the desired accuracy is achieved for frequencies from 0 to $\omega = 0.305\,\omega_n$. For a damping ratio of 0.59 the desired accuracy is achieved for frequencies from 0 to $0.87\,\omega_n$. To make the range of input frequencies of friction force that can be measured accurately as large as possible, the natural frequency must be as high as possible. Therefore, the elastic support must have a large stiffness and the element must have a small mass. However, since the stiffness of the support determines the deflection that is used to measure the friction force, the stiffness

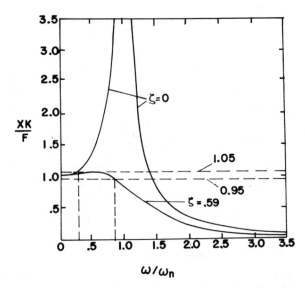

Figure 3.3 Frequency response of a damped second-order system.

can be increased only to a value that results in deflections that can be measured by the detection system.

Of the measurement techniques mentioned, the piezoelectric crystal has the greatest potential for creating a system with a very high natural frequency. These crystals are extremely stiff and have low mass. Therefore, if one of the elements can be mounted on a piezoelectric crystal, the frequency range of accurately measured friction forces will be at a maximum.

One method for measuring the static coefficient of friction uses the inclined plane. In this method one element is the surface of the inclined plane and the other element is a slider placed on the plane. The angle of inclination is increased until the slider slides down the plane. The tangent of the angle at which sliding commences is the static coefficient of friction. The accuracy of this technique depends on the rate at which the angle is changed and the accuracy of the measurement of the angle at which the sliding commences.

4.2 Measurement of Wear

The most common way to measure mass loss from one element is to weigh the element before and after the experiment. The accuracy of this method depends not only on the accuracy of the balance used for weighing, but also on weight gain or loss caused by factors other than the wear process. These factors include loss or gain by corrosion, gain by absorption of moisture, and loss or gain by transfer of material to or from the specimen holder as the specimen is installed and removed for weighing. Compensation for the former two factors is achieved by subjecting a control specimen to the same environment as that of the test specimen. The weight gain or loss of the control specimen is used to adjust the weight of the test specimen. Little can be done to correct for the last factor. The variation caused by it can be calculated from the weight changes resulting from installing and removing specimens several times. This variation can then be added to other measured variations to obtain the overall accuracy of the measurements.

When materials chemically react with the environment while sliding, the reaction products that are chemically bonded to the surface include chemical elements that were not present in the sliding material. Some investigators remove this material by using solvents that dissolve the reaction products but not the original material before weighing the sliding material. For steels, the reaction product is typically alpha iron oxide, which can be dissolved with a 10% ammonium citrate solution [33].

The second common way to determine mass loss is by measuring a dimensional change in the element. This change is then used to calculate the volume lost. A catalog of these calculations for several commonly used element geometries appears in [34]. Two examples in Figure 3.4 show that the measurement h can be used to calculate the wear volume. The wear can be continuously monitored if h is measured continuously by a displacement transducer.

Calculations of wear volume from the dimensional changes of wearing elements are only as accurate as the measurements of the original dimensions and the changes to those dimensions. In addition, changes in the temperature of the sliding elements either caused by frictional heating or changes in the ambient temperature may cause significant changes in the dimensions. Thus the dimensional changes measured during the test may not be solely caused by loss of material.

There are certain sliding systems in which a nonwearing element slides repetitively in a path on the surface of a wearing element. This configuration is used to measure the wear of thin coatings or altered surface layers of materials. A groove is worn into the surface layer or coating. The groove cross-sectional area is measured in several places, and the wear volume is the product of the average area and the length of the wear groove. A stylus surface profiler is often

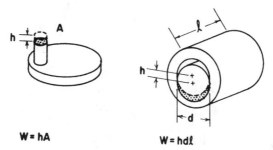

$$W = hA \qquad\qquad W = hd\ell$$

Figure 3.4   Measurement of wear volume.

used for this measurement [35]. The cross-section profile of the groove is measured; then the area of the groove is calculated by numerically integrating the digitized profile data or by measuring with a planimeter the area on a chart recording of the profile.

The accuracy of this method depends on several factors. If the groove is formed both by plastic grooving and the formation of wear particles, then the groove profile will include the groove as well as the plowed material above the original surface. Thus, there are several ways to estimate the groove area, which are illustrated in Figure 3.5. In Figure 3.5a the area is the material that is removed below the original surface regardless of whether some of that material was just plowed out of the groove and still resides at the sides of the groove. In Figure 3.5b the area of the plowed material is subtracted from the area of the groove. The resulting area represents the material that is removed from the surface. In Figure 3.5c the area is measured from the highest points of the plowed material on each side of the track. This last method is useful if maximum roughness of the worn surface is of interest.

The other factors that affect the accuracy of the groove area include the variation of the groove area along its length. It is essential that the variation in groove area along the length is random. If so, then measurements can be made at several random locations used to calculate an average groove area. However, if there are systematic changes in the area because of inhomogeneous properties or anisotropy in properties, then the area measurements should be made to quantify the effects of these on the wear. Finally, the accuracy of the method

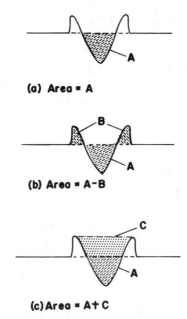

(a) Area = A

(b) Area = A−B

(c) Area = A+C

Figure 3.5   Measurements of groove cross-sectional area.

used to calculate the area from the profile data also affects the accuracy of the area.

Radioactive tracers can be used to determine the wear of components in a complex mechanical system like an automobile engine. The component, such as a cam, is irradiated so that one of its atomic elements is made radioactive. The cam is installed in the engine, which is then run through a duty cycle. The engine oil is monitored with a Geiger counter, and the counts that are recorded by it can be related to the wear of the cam by a calibration procedure. This procedure requires knowledge of the weight percent of the radioactive element in the cam material and the half-life of the radioactive element. The major advantage of this system is that very small amounts of wear can be detected. Consequently, relatively short tests can be run under normal operating conditions, and wear can be measured. The major disadvantage of this method is the requirement to handle and store radioactive materials.

The radiotracer technique can be used to measure the transfer from one sliding element to another. This technique was used to measure the transfer from a polymer pin that contained radioactive chlorine to a metal disk [36]. The amount of activity detected on the disk after the experiment was related to the mass of polymer transferred by a calibration. While this technique could not be used to obtain a continuous measurement of wear, other methods of surface chemical analysis were used in this way.

Basic studies on the transfer of materials from one sliding element to another have utilized electron spectroscopy for chemical analysis (ESCA) and Auger electron spectroscopy (AES) to detect transferred material. Because these techniques require a vacuum the experiments are performed in vacuum chambers. In one configuration the transfer from a pin to a rotating disk was detected by an AES detector that was positioned to analyze the wear track diametrically opposite the pin [37]. The resultant spectra can be used to determine the relative amounts of transferred atomic elements.

5 COMPONENTS OF LABORATORY DEVICES (TRIBOMETERS)

Most laboratory devices have the following characteristics: (1) geometry of contacting elements, (2) normal load application, (3) relative motion, and (4) component and measurement interactions.

5.1 Element Geometry

The geometries of the elements in tribological systems can be grouped into three categories: *point, line,* and *area contact* based on the contact geometry when the normal load is zero, as shown in Figure 3.6. Point contact has the advantage that no alignment is required. Line and area contact require perfect alignment or the contact becomes a point contact. When a load is applied these geometric contacts become area contacts as the materials deform. A catalog of several common geometric configurations is given in [5].

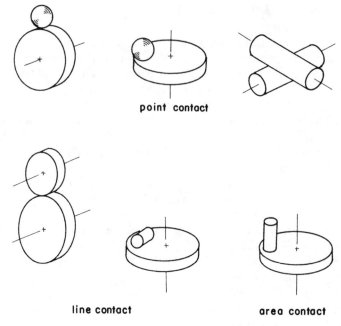

point contact

line contact   area contact

Figure 3.6   Contact geometries.

5.2   Normal Load

The normal load can be applied by dead weight either directly or through a lever that forces one element onto the other. Pneumatic or hydraulic cylinders are also used, the latter when very large loads are required. Other methods of applying the normal load are magnetically, through a spring, or through the expansion of the constrained members that are heated. Since both wear and the friction force vary monotonically with the normal load, it is desirable to keep the normal load constant during the test. A constant normal load can be obtained with all of the preceding methods except for the spring and thermal expansion. As the elements wear and the length of the elements change, the force exerted by the spring and the expansion of the elements changes.

If the normal load is applied by dead weight acting on a stationary element and high sliding speed is required, the run out of the moving element can impart vertical accelerations to the dead weight. The force required for these accelerations will cause fluctuations in the normal load. A pneumatic loading system uses pressure against a piston to develop the normal load. In this system the mass accelerated by the run out is smaller than that required to obtain the same normal load by dead weight. Thus the pneumatic loading system is used to reduce fluctuations in normal load when high rotational speeds are required.

If the normal load is applied through a lever, the pivot of the lever must lie on the projection of the friction force vector. If this is not done, then the moment created by the friction force about the pivot will affect the normal load between the elements as shown in Figure 3.7. The equation in the figure shows how the proportions of the lever and the offset of the pivot from the line of action of the friction force affect the magnitude of the normal load. Note that this effect can either increase or decrease the normal load depending on the direction of rotation of the cylinder and the location of the pivot.

5.3 Relative Motion and Contact Ratio

The relative motion between the elements can either be reciprocating or unidirectional. In reciprocating sliding the sliding velocity must reverse direction twice per cycle. Special mechanisms or control systems can be used to maintain constant velocity over most of each stroke. At the point when the velocity reverses, the relative velocity between the elements is zero. This condition could initiate stick–slip friction, which is one reason why reciprocating motion is used to study this phenomenon. The advantage of unidirectional sliding is that it can have constant velocity for the entire test.

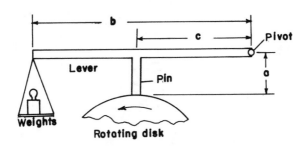

(a) Loading lever

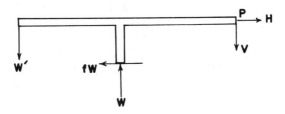

(b) Freebody diagram of lever

$$\sum \text{moments about } P = 0$$
$$W'b - Wc - fWa = 0$$
$$W = \frac{W'b}{c + fa}$$

Figure 3.7   Normal load applied by a lever.

The geometry of the elements and the type of motion determine the contact ratio for each element [38]. The contact ratio is the ratio of the apparent contact area to the total area on an element that contacts the other element during the test. The contact ratio for three geometries is illustrated in Figure 3.8a–c. In part (a) the surface area of each washer is always in contact with the other. In this case the contact ratio is 1 for each washer. In part (b) the area at the end of the pin is always in contact with the surface of the disk. However, an area on the disk (equal to the pin end area) is only in contact with the pin once per revolution. The contact ratio for the pin is 1 but the ratio for the disk is $r/2R$. In part (c) both elements have contact ratios of less than 1.

Elements with a contact ratio of 1 are in continuous contact with the other element. Therefore, they experience the continuous application of the friction force and the heat generated by dissipation of the frictional work. Macroscopically the element is in a steady-state sliding mode. For an element with a contact ratio of less than 1 the areas that contact the other element do so intermittently and therefore see transient conditions of force and heat input. If one is using a lab test to simulate a system, the contact ratios of the elements in the lab test should duplicate those for the elements in the system.

(a) Contact ratio for A=1, B=1

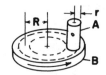

(b) Contact ratio for A= 1.0, B= r/2R

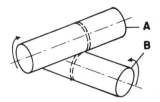

(c) Contact ratio for A<1.0, B<1.0

Figure 3.8   Contact ratio.

5.4   Component and Measurement Interactions

It is most common to have one element moving and one stationary to achieve a desired relative motion. It is convenient to apply the normal load to and measure the friction force and wear continuously on the stationary member. This convenience is illustrated next. If the normal load is applied to the moving element, then a bearing must be used to transmit the load from dead weight or hydraulic or pneumatic pressure to the moving element. The bearing will also contribute to the friction in the system and may also affect the friction measurement. Application of the normal load through the stationary member avoids this problem.

If the friction force is measured on the moving member, the information must be transmitted from the moving member to the stationary recording device. For example, if strain gauges are used to measure the strain in a rotating shaft caused by friction force on a disk or cylinder, slip rings must be used to conduct the current to and from the gauges. Because of the contact resistance of the slip ring-brush interface the electrical noise sets a limit on the sensitivity of the measurement. This limitation does not exist if the friction measurement is made on the stationary member because the sensor can be hard wired to the recorder.

For systems that have reciprocating motion where the velocity and acceleration of one member changes periodically, the friction force should be measured on the stationary member for the most accuracy. If it is measured on the moving member, the inertia force (mass of the element between the point of measurement and the location of the friction force times its acceleration) is also included in the the measurement of the friction force. Because the acceleration varies as the square of the frequency in systems being moved with harmonic motion, the influence of the inertia force on the measured friction force increases at higher frequencies.

Continuous measurements of the change in length of a pin or height of a washer are made more easily on the stationary member. Continuous wear measurements by radiotracers can be made on moving or stationary members with equal ease. If wear is to be measured by periodically stopping the apparatus and weighing or measuring an element, both elements should be easily removed and replaced in the apparatus.

6   TEST PROCEDURES

In a tribological experiment there are periods of transient or steady-state behavior or both. Continuous or periodic measurements must be made to determine when the periods of steady state exist. Transient periods always exist at the start of an experiment. These transients are caused by many factors.

6.1   Transient Behavior

In systems where the theoretical contact at no load is a line or an area, misalignment of the elements produces a theoretical point contact. When the normal load is applied, an area will be developed to support the load. If the stress in the area exceeds the yield strength of the material, then plastic flow will occur. The commencement of motion introduces a tangential force that causes the contact area to increase. A possible consequence of the plastic deformation of the material is wear. Initially, a very rapid wear can occur, which results in an increase in the area of contact. The concomitant reduction in the stress level causes the wear rate to decrease. This transient continues until the system achieves a contact geometry where the nominal stress level (based on the apparent area of contact) is below the yield strength of the material and the real area of contact is less than the apparent area. This process is often called *running-in.*

Even if the initial alignment of the elements is perfect for the theoretical line and area contacts, the friction force can deflect the elements and cause the line and area contacts to become point contacts. This condition is particularly troublesome in the pin-on-disk configuration when the pin is made of a low modulus material such as a plastic.

Initial transients can also be caused by the dynamic response of the system as it is put into motion. Experiments can be started in one of two ways. The elements can be statically loaded and then put into motion or the motion can be started before the normal load is applied. Either method produces transients; however, the transients are usually very short compared to the length of the experiment and therefore are of little consequence unless either procedure produces unusually high loads. For example, in a lubricated experiment, when the system is at rest, the lubricant penetrates the apparent area of contact. When the motion is started, this lubricant film must be sheared at a high rate, which produces a force large enough to destroy the support for one of the elements. This mode of failure can occur to the supports for magnetic heads that are resting on lubricated magnetic rigid disk surfaces during start-up [39].

Transients can also occur as a result of frictional heating. The influence of thermal transients on friction and wear depends on the temperatures and temperature gradients produced and on the sensitivity of the structure and properties of the materials to these thermal conditions. The duration of the transients can have a large range. In the pin-on-disk configuration an area in the sliding track on the disk experiences short thermal transients each time it passes under the pin. The temperature rises while it is in contact with the pin and then falls when it is not in contact. If the time constant for the exponential rise and fall of the temperature is short compared with the time between passes under the pin, then the temperature can return to ambient after each contact. However, if the time constant is of the same magnitude as the time between contacts, then the disk temperature will rise until an equilibrium is reached between the heat gained during the pin contact and the subsequent heat lost to the environment. This is a long-duration thermal transient and it also increases the pin temperature.

Thermal inputs will cause the dimensions of the system to expand or contract in response to temperature changes. Thus, if wear is being measured by a change in the length of a pin, the thermal expansion of the pin causes an error in this measurement. Thermal expansion can cause changes in bearing clearances or preloads in a drive system that can change friction torques. If these torques are also being measured by the friction measuring system (such as the measurement of the twist of a shaft by strain gauges), errors in the friction result.

In experiments where an initial transient wear is followed by steady-state wear (i.e., the wear rate is constant), the steady-state wear rate must be calculated from the wear measurements made while the wear rate is constant. In Figure 3.9 wear of a pin versus time curves are shown for three values of surface roughness of a disk. If only one wear measurement is made at the same time for each disk and the wear rate is calculated to be the wear volume divided by time, then all the calculated wear rates will be higher than the steady-state wear rates. The error will be largest for the surface that produces the longest duration transient period. Wear volumes must be measured at different times to determine the duration of the transient, and only those values measured during steady state can be used to calculate the wear rate.

An experiment where the transient conditions are the primary concern is the test to measure the pressure-velocity (PV) limit of a polymeric material [40]. The PV limit is a measure of the maximum energy input to a material for which it will achieve a steady-state temperature. The PV limit is based on the assumption that the wear volume W is proportional to the normal load L and the sliding distance S (2).

$$W = KLS \tag{2}$$

where K is a proportionality constant and S is the product of velocity V and time t. From Figure 3.10, W is hA. Thus, (2) can be rearranged to (3):

$$h/t = K(L/A)V \tag{3}$$

where V is the average sliding velocity over the area A. The pressure P over area A is L/A, and (3) can be written in terms of the product PV (4):

$$h/t = KPV \tag{4}$$

In this experiment either of two procedures can be used. For a constant sliding velocity a load is applied, and the temperature of one of the sliding elements is measured. When the temperature reaches steady state, the load is increased. This stepwise increase in load is continued until the temperature rises rapidly and high wear results. The product PV prior to this unstable condition is

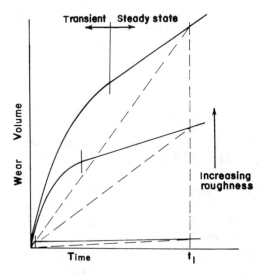

Figure 3.9 Transient and steady-state wear.

called the *PV* limit. In the second procedure, the load is fixed and the velocity is increased in a stepwise manner.

In one sense friction and wear are always transient phenomena especially if these phenomena are viewed at the microscopic level over very short time spans. However, if instantaneous measurements of friction and wear are averaged over longer time intervals, these average values can be constant. These constant long-term averages are usually called *steady-state conditions*. In general, investigators are concerned with long-term averages and how these averages change as a result of time, operating conditions, materials, and environment.

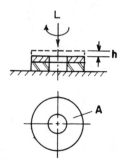

Figure 3.10   Linear wear in a thrust washer test.

6.2   Friction Measurement

Because the friction force is a time-varying phenomenon that contains a spectrum of frequencies, the frequency response of the measuring technique will affect the measured value as noted in Section 4.1. The objective of a measuring technique is to measure the amplitude of the friction force over as wide a range of frequencies as possible without distortion. For frequencies outside these limits the values are attenuated or magnified. Because the investigator may not be interested in the microscopic friction events that cause the time dependent variations, the electronic signal from the force transducer can be passed through an electronic low-pass filter that attenuates all but the lowest of frequencies. In this way the steady-state value of the friction is extracted from the fluctuating signal. However, in a more common procedure the mean value of the fluctuations is approximated from a chart recording of the friction force.

Unfortunately, fluctuations in the friction force can be caused by factors other than point-to-point variations in surface roughness and adhesion. One of the more common factors is caused by fluctuations in normal load. Normal load fluctuations are usually caused by inertial forces as described in Section 5.2. Because the inertial force frequency is a function of the speed of rotation of one of the elements, this source of variation can be identified in the frequency spectrum of the friction force variation. The spectrum usually contains components that are equal to or multiples of the rotational frequency. Techniques such as fast Fourier transform (FFT) can be used to obtain such spectra [41]. Estimates of these components of friction variation can also be made from chart recordings made at high chart speeds so that periodic components of the friction force can be easily seen. These fluctuations can also be filtered out by electronic filters that do not pass the rotational frequencies of the element.

6.3   Wear Measurement

The most important parameter used to quantify wear is the wear rate. Wear rate is defined as the volume (or mass) of material lost per unit of time. Since most measurements of wear are made over a period of time where the relative sliding velocity is constant, the sliding distance is proportional to time. Therefore, the more common expression for wear rate is *volume lost per unit sliding distance*. This form of wear rate is also used in reciprocating sliding experiments where the velocity is variable.

The wear rate can be divided by the normal load to give a quantity called the *specific wear rate*; if the wear rate is proportional to the normal load, the specific wear rate can be used to compare wear data measured at different normal loads. If the wear is not proportional to load, say

$$W = KL^{0.9}S \tag{5}$$

then the specific wear rate is

$$\frac{W}{LS} = \frac{K}{L^{0.1}} \tag{6}$$

Here the specific wear rate is dependent on the load. When the wear is not proportional to load, then the comparisons can be made only at a specific load.

If the specific wear rate is multiplied by the hardness of the wearing material, the result is the *wear coefficient*, which is dimensionless when the wear and the hardness are expressed in units of volume and units of force per unit area, respectively.

$$\frac{WH}{LS} = K \tag{7}$$

An extensive description of wear coefficients and typical values is found in [23]. These parameters; that is, wear rate, specific wear rate, and wear coefficient, are used in the literature.

In tribological systems where the contact ratio for one or both of the elements is less than 1, an alternate form of wear rate is used. Because any point on the path of contact on the element contacts the mating element only once per revolution, the distance in the wear rate expression is replaced by the number of revolutions times the circumference of the contact path. The wear volume is replaced by the groove cross-sectional area times the circumference. Thus the wear rate becomes the average groove cross-sectional area per revolution of the element [35]. Techniques that measure wear continuously (see Section 4.2) provide data from which instantaneous wear rates can be calculated. For techniques that measure the change in a dimension of an element, such continuous wear measurements may be confounded with effects of run out of an element or dimensional changes caused by thermal expansion. Fluctuations of the instantaneous wear rates can be analyzed by using the techniques described in Section 6.1 for the analysis of friction force variations.

Two procedures are used for making periodic measurements of wear. In the first, the system is run for a period of time after which the elements are removed from the system for the wear measurement. Then they are replaced in the system, and the test is continued until it is stopped for a second measurement. This procedure continues until several data points are obtained, and it has the advantage that the same two elements are used for the successive wear measurements. There are several disadvantages of this procedure. When the elements are replaced in the system they may not be in the same orientation as they were before they were removed. Therefore, some run-in wear will occur to reestablish conformity between the surfaces. (One apparatus avoids this pro-blem by having a spherical element mounted in an arm that can pivot so that the wear scar on the sphere can be measured in a microscope without removing the sphere from the arm.) When the experiment is stopped for a wear measurement the temperature of the specimen changes. Thus, when the test is restarted, a thermal transient will occur as the system tries to establish a steady state.

In the second procedure two elements are slid continuously until a wear measurement is to be made. The elements are not used for further testing. To obtain the wear for various sliding distances, different elements are used for each sliding distance. The advantage of this procedure is that for long sliding distances the temperatures can reach steady state without being disrupted by intermediate stoppages for wear measurements. The problem of realigning the elements when they are returned to the apparatus is eliminated. The disadvantage is that different elements are used for each measurement, and the tests require more materials and longer times.

6.4   Abrasive Wear Tests

Some systems (magnetic recording media) are designed for total wear on the order of the original surface finish, while others (clutches and brakes) are designed for much larger total wear. The environments in which these systems operate sometimes contain small particles that pass through the contact area between the elements. If the particles are harder than either of the two elements, then abrasive wear of these elements can occur. In some systems, such as the tribological contact between a mining machine and a mineral or ore, abrasive wear occurs continuously. When designers select materials for such systems, the abrasive wear resistance of the materials is a criterion. Several tests were developed to measure abrasion resistance of materials (abrasion resistance is the reciprocal of the wear rate in an abrasion experiment).

These tests are used to expose a material to moving abrasive particles that are forced against the surface of the material. In one test (the ASTM standard practice for *Conducting Dry Sand/Rubber Wheel Abrasion Tests*, G65) the abrasive particles are fed into an interface between the material to be tested and an elastomer. The particles are dragged through the interface by the relative motion between the material and the elastomer. The normal load between the material and the elastomer causes the particles to be partially anchored by the elastomer while they abrade the material. When the particles leave the interface they are ejected from the elastomer surface. In one configuration the material specimen is stationary, the elastomer is a rotating wheel, and the abrasives are fed by gravity into the interface. In another configuration, the material specimen is moved in a reciprocating motion on a rubber pad in the presence of an abrasive slurry. The specimen is lifted in each stroke to permit fresh slurry to enter the contact region. The ASTM test method G75, *Slurry Abrasivity by Miller Number*, utilizes this configuraton to test abrasiveness of slurries. However, if a standard slurry is used, the test can be used to measure abrasive resistance of materials.

In another class of tests, the abrasives are fixed to a paper or cloth backing. The material specimens are either stationary or moving slowly, and the abrasive paper is moved rapidly over the specimen. The abrasive material can pass by the specimen once or it can have multiple passes on the specimen. For single passes over the specimen, the problem of wear debris clogging the abrasive surface is eliminated at the expense of the requirement for enough abrasive surface to cause measurable wear of the specimen. In ASTM standard test method D1242, *Resistance of Plastic Materials to Abrasion*, a roll of abrasive paper 50 m long is pulled past the specimen. In another form of the test we use an abrasive paper disk, which causes the specimen pin to move radially while the disk spins to create a spiral path on the disk. Tests that have multiple passes on the abrasive surface require much less abrasive surface than single-pass tests. However, in addition to debris clogging the abrasive surface, the abrasives can wear and become duller as the test proceeds. Comminution and dulling of the abrasives are possible in the test involving loose abrasives if they are recirculated through the interface during the test.

6.5   Wear Debris

One of the consequences of a wear test is the generation of wear debris. If the debris is trapped in the interface or remains in the wear track, it can affect subsequent wear [42, 43]. If the debris consists of oxides, which are softer than the element surfaces, their presence can serve to distribute the normal load over a larger area, which will reduce stresses and subsequent wear. The softer debris can also act as a lubricant. If the debris consists of workhardened material or oxidized material that is harder than the surface, then the debris can be abrasive to the surfaces and increase the wear.

Some systems are oriented so that gravity forces tend to remove the debris from the wear track (e.g., disks rotating about horizontal axes). This arrangement can be convenient if the debris is to be collected and analyzed for composition, morphology, and size distribution.

The geometry of the elements affects the probability of the debris in the wear track reentering the interface. If the elements are shaped so that the debris approaching the interface is wedged between the two elements, then the probability is high that the debris will be drawn into the interface. An example of this geometry is a sphere sliding on a disk. On the other hand, the geometry may cause the debris to be pushed to the side of the wear track. A cylindrical pin with a vertical axis sliding on a horizontal disk tends to push the debris aside.

In some tests the debris is removed from the wear track by air jets to study the effect of the debris on the wear process. A study of graphite seals shows that removal of the debris increases the wear rate by an order of magnitude [44]. In a study of the wear of siloxane-modified epoxies the wear increased when the debris was removed [45].

7   STATISTICAL CONSIDERATIONS

Measurements of friction force and wear volume are subject to the same errors as are all measurements. In addition, the phenomena being measured are inherently random. Therefore, experiments must be replicated so that mean values and variances of the data can be calculated. Once they are calculated then the means and variation can be reported in several ways as indicated in Table 3.2.

If mean values are plotted on graphs, error bars should be used to indicate

Table 3.2 Methods of Reporting Means and Variability

Parameters	Comments
Mean only	No indication of variability or number of data points
Mean, standard deviation	Variability given, but no indication of the number of data points
Mean, range	Range gives no indication of dispersion of data within range; it is easy to calculate
Mean, standard deviation, number of data points	Sufficient information to calculate confidence intervals if desired
Mean, confidence intervals	Most informative; confidence intervals are calculated from standard deviation and number of data points

variability. The figure legend should clearly identify the error bars; for example, range, standard deviation, or 95% confidence intervals. Because the meaning of confidence intervals is not as well understood as standard deviation and range, a brief explanation follows.

7.1   Confidence Intervals

When replicate measures of wear are made, the calculated average wear will seldom be the same as the true average of wear. It would be desirable to state with some confidence an interval that will bracket the true average wear. If several samples of replicates are made and an interval is calculated about the average wear in each sample, it will be found that a certain percentage of the intervals, say 95%, contains the true average wear. Therefore, it will be possible to state that the confidence that any one interval contained the true average is 95%. This interval can be calculated for a given sample of replicate measurements of wear using the mean and standard deviations with the following equation [46]:

$$\bar{X} - \frac{ts}{\sqrt{n}} < \mu < \bar{X} + \frac{ts}{\sqrt{n}} \tag{8}$$

where $\bar{X}$ is the sample average, s is the sample standard deviation, n is the number of replicates in the sample, t is the value of Student's-t distribution at a confidence level of $1 - \alpha$ for $n - 1$ degrees of freedom, α is the probability that $t > |t_{\alpha/2}|$ (see Figure 3.11), and μ is the true average value.

Wear rate is often determined from a series of wear volume measurements taken at different sliding distances. It is assumed that the wear volume is proportional to the sliding distance, and a statistical process called *linear regression* is used to determine this relationship. A straight line $W = a + bS$ is

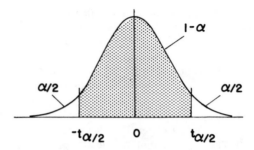

Figure 3.11   Student's-t distribution.

fitted to the wear volume W versus distance S data using the least-squares principle. The slope b of this line is the wear rate, and because of the randomness of the wear data the slope has a variation. Equations (9) and (10) indicate how to calculate the slope and its variation [47]. The value of b is given by

$$b = \frac{n(\Sigma WS) - \Sigma W \Sigma S}{n(\Sigma W^2) - (\Sigma S)^2} \tag{9}$$

where n equals the number of pairs of W and S values. The confidence intervals for b are

$$b - \frac{ts_e}{\sqrt{\dfrac{n(\Sigma S^2) - (\Sigma S)^2}{n}}} < \beta < b + \frac{ts_e}{\sqrt{\dfrac{n(\Sigma S^2) - (\Sigma S)^2}{n}}} \tag{10}$$

where

$$s_e = \sqrt{\frac{\Sigma W^2 - a\Sigma W - b(\Sigma WS)}{n - 2}}$$

t is the Student's-t distribution parameter for a confidence level of $1 - \alpha$ and $n - 2$ degrees of freedom, β is the true value of slope, and

$$a = \frac{\Sigma W \Sigma S^2 - \Sigma W \Sigma WS}{n(\Sigma W^2) - (\Sigma S)^2}$$

7.2   Comparison of Means; Hypothesis Testing

When wear is measured as a function of a variable, say material composition, a statistical procedure called *hypothesis testing* can be used to determine whether the wear of material A is different from that of material B. In this procedure, the hypothesis is made that there is no difference between the mean values of wear for A and B. Then a statistical comparison is made to determine whether the hypothesis is true. If it is true, then a statement can be made that there is no difference between the wear of A and that of B with a confidence level of, for example, 95%.The test [48] can be stated as follows:

$$|X_A - X_B| > t' \sqrt{\frac{s_A^2}{n_A} + \frac{s_B^2}{n_B}} \tag{11}$$

where X_A, X_B are the sample averages; n_A, n_B are the number of replicates per sample; s_A, s_B are the sample standard deviations; t' is the value of Student's-t distribution for f degrees of freedom and a confidence level of $(1 - \alpha)$; and

$$f = \frac{s_A^2/n_A + s_B^2/n_B}{\dfrac{(s_A^2/n_A)^2}{n_A - 1} + \dfrac{(s_B^2/n_B)^2}{n_B - 1}} - 2$$

If the inequality is satisfied, the statement can be made that the mean values of wear of A and B are not equal with a confidence of, for example, 95%.

7.3   Factorial Experiments

One common method of experimentation is to fix all variables except one and then determine the effect of that variable on the friction and wear. However, this method does not guarantee that the variations of the friction and wear found will be the same if the values of the fixed variables were chosen differently. A technique called *factorial design* can be used to determine whether such interactions between variables are significant. An example of interactions between two variables is illustrated in Figure 3.12. In Figure 3.12a the parallel lines indicate that the change in film life, which occurs with the change in load, is the same at each value of the film thickness. Hence, there is no interaction between load and film thickness on film life. In Figure 3.12b the converging lines indicate that at the low amplitude of motion an increase in frequency causes a decrease in film life, while at high amplitude an increase in frequency causes an increase in film life. Thus, there is an interaction between amplitude and frequency on film life.

In factorial design the variables that are expected to affect the friction and wear are given two or more values. The number of tests then becomes the total number of combinations of the variables and their levels. For example, if the variables are normal load (5 and 10 N), sliding speed (0.1, 0.5, and 1 m/s), and relative humidity (10 and 90%), then the total number of combinations will be $(2)(3)(2) = 12$. The 12 tests can be replicated two or more times to obtain data that can be used to estimate the experimental error caused by randomness of the data. The analysis of the results of the experiment gives the mean values of the

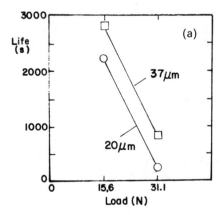

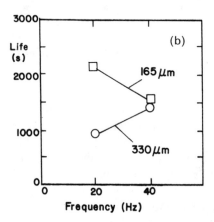

Figure 3.12 Interactions between factors [49]. (a) Fretting life versus load and film thickness. (b) Fretting life versus frequency and amplitude.

friction and wear for each level of each variable and the mean values of the interactions and identifies the variables and the interactions between the variables that are statistically significant at a chosen confidence level [50].

If the number of variables is large, say 6, and each is tested at two levels, then 2^6 or 64 combinations results. In this case the interactions between combinations of 4, 5, and 6 variables will be used to estimate the experimental error because such higher order interactions are not likely to be significant. If the number of combinations is very large, say 10 variables at 2 levels each (1024 combinations), a technique called *fractional factorial* design can be used. In this technique a number of combinations less than the 1024 is selected in such a way that higher order interactions, which are not likely to be significant, are confounded with the estimates of the main effects of the variables [51].

Another important aspect of experiment design is to prevent uncontrolled variables from influencing the results. For example, if the wear of three different materials is to be measured over a period of 6 months in a laboratory that is not air-conditioned, the humidity can be low in the winter and high in the summer. If the test is designed so that material A is tested in the winter, material B in the spring, and material C in the summer it will not be possible to determine whether the differences observed in the wear rates were caused by the differences in materials or in humidity. To design the experiment properly the testing order of the three materials should be randomized so that each material is tested in each humidity condition [52].

7.4   Statistical Versus Practical Significance

While statistical procedures can be the basis for decisions on the significance of the differences between measured variables such as friction or wear, it must be noted that there is a distinction between a difference that is statistically significant and one that is of practical importance. Because the tests of significance depend on the number of replicates of a measurement in a sample it is always possible for the researcher to take such a large sample that any difference is significant or to take such a small sample that no difference is significant.

Likewise, the statistical procedure depends on the level of significance selected by the researcher. The consequence of the level of significance is its effect on two types of incorrect conclusions from the statistical tests. These errors are (1) rejecting a true hypothesis (called a *Type I error*) and (2) failing to reject a false hypothesis (called a *Type II error*). It is the researcher's responsibility to evaluate the consequences of either error on the practical significance of the conclusions of the experiment. Then the researcher can choose the level of significance that minimizes the practical consequences of an erroneous conclusion. For more insight into this decision process the reader is referred to expositions on inferential statistics and hypothesis testing [53].

8   CONCLUSIONS

Friction and wear phenomena depend on a multitude of interactive and inherently random factors. Consequently, if the values of friction and wear are desired for the prediction of performance or durability, they must be measured in the exact conditions of the application. Accelerated tests for wear measurement are valid when the wear varies linearly with the factors that are used to increase the wear rate, such as normal load or sliding velocity, and the wear mechanism is the same in the accelerated and normal application. Laboratory tests can be useful for screening materials for a tribological application when the predominant mechanisms of friction and wear are the same in the test and the applications. Some standard tests were developed that are useful in ranking the wear of materials when subjected to abrasive or sliding (nonabrasive) wear.

Because of the random nature of friction and wear, replicate measurements must be made and the results must be expressed in terms of mean values and a measure of variability. The statistical procedure called *hypothesis testing* must be used to determine whether differences in mean vaues of friction and wear are significant. Confidence levels and sample sizes must be chosen to maximize the practical significance of the results of the experiments.

9   APPENDIX

Tables 3.3 and 3.4 list tests for friction and wear that were published by ASTM.

Table 3.3   ASTM Friction Tests[a]

Code	Materials[b]	Applications, Device
B460	Sintered MET	Dry, heated brake drum
B461	Sintered MET	Lubricated, heated brake drum
B526	Sintered MET	Dry clutch breakaway torque
C808	Graphite	Seals, data reporting
D202	Paper	Insulation, inclined plane
D1894	Plastics	Sheet, films; five devices
D2047	Floor polish	Floor coatings
D2394	Wood	Finish flooring, weighted slider
D2534	Wax coating	Self-mated, friction sled
D2714	MET CER PLS	Block-on-ring device
D3028	Plastics	Solid, sheet, film; pendulum
D3108	Yarn-metal	Textile machines, wrapped on drum
D3247	Fiberboard	Corregated, solid; horizontal plane
D3248	Fiberboard	Corregated, solid; inclined plane
D3334	Fabrics	Inclined plane
D3412	Yarn–yarn	Twisted strand, hung on capstan
D4103	Floor polish	Flooring, surface preparation
E303	Pavement	Field test, portable pendulum
E510	Pavement	Small torque motor bench test
E670	Pavement	Tire side force
E707	Rubber, fabric	Tire skidding, pendulum device
F489	Rubber, leather	Shoe holes, heels
F524	Carbon paper	Tilting plane device
F695	Rubber, leather	Shoe soles, heels; data evaluation
F732	MET CER PLS	Protheses, reciprocating pin-on-flat

[a]Adapted, with permission, from P. J. Blau, *Friction and Wear Transitions of Materials: Break-in, Run-in, Wear-in*, Noyes Publications, Park Ridge, NJ, 1989, p. 33 [54].
[b]MET, metals, alloys; CER, ceramics, glasses; PLS, plastics.

Table 3.4   ASTM Wear Tests[a]

Code	Materials[b]	Applications, Device
C131	Mineral aggregate	Crushing resistance, ball mill
C418	Concrete	Sand blasting, air driven sand
C448	Ceramics	Porcelain enamel, abrasive drum
C501	Ceramics	Tile, abrasive drum on tile
C585	Mineral aggregate	Crushing resistance
C704	Ceramic	Refractory brick, jet erosion
C779	Concrete	Abrasion, sliding, impact
C808	C-graphite	Seals, data reporting
C944	Concrete	Rotary cutter, drill press
D658	Organic coatings	Paint, lacquer, varnish; jet abraded
D968	Organic coatings	Paint, lacquer; falling sand or SiC
D1242	Plastics	Floor tile, one-pass abrasive
D1395	Organic coatings	Floor coverings
D1630	Rubber	Shoe soles, heels; abrasive drum
D2714	MET CER PLS	Sliding wear, ring-on-block
D3181	Textiles	Clothing wear, human subjects
D3702	Metals	Self-lubricated, thrust washer
D3884	Textiles	Abrasive drum on moving fabric
D3885	Textiles	Woven fabrics, rubbing and flexing
D4157	Textiles	Abrasion, oscillating cylinder
D4158	Textiles	Abrasion, rotary rubbing
G6	Insulation	Pipe coating, revolving drum
G32	MET CER PLS	Cavitation, vibration in liquid
G56	Fabric	Inked ribbon, ribbon wrapped drum
G65	Metals	Abrasion, sand-rubber wheel
G73	MET PLS	Erosion, drops impact spinning specimen
G75	Metals	Slurry abrasivity, reciprocating lap
G76	MET CER PLS	Erosion, particles against flat
G77	MET CER PLS	Sliding wear, block-on-ring
G81	Metals	Jaw crusher, angled plates
G83	Metals	Sliding wear, crossed cylinders
F510	Organic coatings	Floor coverings, abrasive drum

[a]Adapted, with permission, from P. J. Blau, *Friction and Wear Transitions of Materials; Break-in, Run-in, Wear-in*, Noyes Publications, Park Ridge, NJ, 1989, pp. 189–190. [54].
[b]MET, metals, alloys; CER, ceramics, glasses; PLS, plastics.

References

1. J. Halling, *Principles of Tribology*, Macmillan, London, 1975, p. 5.
2. F. P. Bowden and D. Tabor, *The Friction and Lubrication of Solids*, Part II, Oxford University Press, London, 1964, pp. 502–516.
3. E. Rabinowicz, *Friction and Wear of Materials*, Wiley, New York, 1965, p. 110.

4. R. L. Pontious, *Lubr. Eng.*, **15**, 110 (1959).

5. R. D. Brown, "Test Methods," in F. F. Ling, E. E. Klauss, and R. S. Fein, Eds., *Boundary Lubrication*, American Society of Mechanical Engineers, New York, 1969, pp. 241–292.

6. M. B. Peterson, "Design Considerations for Effective Wear Control," in M. B. Peterson and W. O. Winer, Eds., *Wear Control Handbook*, American Society of Mechanical Engineers, New York, 1980, pp. 431–438.

7. H. Czichos, *Tribology*, Elsevier, Amsterdam, 1978, pp. 246–286.

8. Society of Tribologists and Lubrication Engineers (formerly American Society of Lubrication Engineers), *Friction and Wear Devices*, ASLE SP-4, 2nd ed., Society of Tribologists and Lubrication Engineers, Park Ridge, IL, 1976.

9. H. Czichos, *Tribology*, Elsevier, Amsterdam, 1978, pp. 58–60.

10. F. P. Bowden and D. Tabor, *The Friction and Lubrication of Solids*, Oxford University Press, London, 1950, pp. 90–98.

11. K. L. Johnson, "Aspects of Friction," in D. Dowson, C. W. Taylor, M. Godet, and D. Berthe, Eds., *Friction and Traction*, Westbury House, Guildford, England, 1981, pp. 3–12.

12. J. Halling, *Principles of Tribology*, Macmillan, London, 1975, p. 78.

13. F. P. Bowden and D. Tabor, *The Friction and Lubrication of Solids*, Part II, Oxford University Press, London, 1964, p. 79.

14. T. Simkins, *Lubr. Eng.*, **23**, 26 (1967).

15. A. W. J. deGee and G. W. Rowe, "Glossary of Terms and Definitions in the Field of Friction, Wear, and Lubrication; Tribology," in M. B. Peterson and W. O. Winer, Eds., *Wear Control Handbook*, American Society of Mechanical Engineers, New York, 1980, pp. 1143–1201.

16. T. O. Mulhearn and L. E. Samuels, *Wear*, **5**, 478 (1962).

17. M. Sawa and D. A. Rigney, "Sliding Behavior of Dual Phase Steels in Vacuum and Air," in K. C. Ludema, Ed., *Wear of Materials 1987*, American Society of Mechanical Engineers, New York, 1987, pp. 231–244.

18. N. Suh, *Tribophysics*, Prentice Hall, Englewood Cliffs, NJ, 1986, pp. 199–209.

19. T. F. J. Quinn, *Tribol. Int.*, **16**, 257–271 and 305–315 (1983).

20. R. B. Waterhouse, "Fretting," in D. Scott, Ed., *Treatise on Materials Science and Technology*, *13*, *Wear*, Academic, New York, 1979, pp. 259–285.

21. F. P. Bowden and D. Tabor, *The Friction and Lubrication of Solids*, Part II, Oxford University Press, London, 1964, pp. 52–78.

22. E. E. Bisson and D. H. Buckley, "Coefficients of Friction," in R. E. Bolz and G. L. Tuve, Eds., *Handbook of Tables for Applied Engineering Science*, The Chemical Rubber Company, Cleveland, OH, 1970, pp. 498–502.

23. E. Rabinowicz, "Wear Coefficients-Metals," in M. B. Peterson and W. O. Winer, Eds., *Wear Control Handbook*, American Society of Mechanical Engineers, New York, 1980, pp. 475–506.

24. K. C. Ludema, *Lub. Eng.*, **44**, 500 (1988).

25. N. S. Eiss Jr., and J. R. Potter III, "Fatigue Wear of Polymers," in L-H. Lee, Ed., *Polymer Wear and Its Control*, ACS Symposium Series 287, American Chemical Society, Washington, DC, 1985, pp. 60–66.

26. J. K. Lancaster, *Tribology*, **6**, 219 (1973).

27. H. Czichos, *Tribology*, Elsevier, Amsterdam, 1978, pp. 264–271.

28. Y. Kawakubo, H. Ishhara, Z. Tsutsumi, and J. Shimizu, "Spherical Pin Sliding Test on Coated Magnetic Recording Disks," in B. Bhushan and N. S. Eiss Jr., Eds., Tribology and Mechanics of Magnetic Storage Systems, Vol. III, Special Publication SP-21, Society of Tribologists and Lubrication Engineers (formerly American Society of Lubrication Engineers), Park Ridge, IL, 1986, pp. 118–124.

29. E. O. Doebelin, *Measurement Systems Application and Design*, 3rd ed., McGraw-Hill, New York, 1983, pp. 211–300.

30. J. Halling, *Principles of Tribology*, Macmillan, London, 1975, pp. 147–173.

31. K. G. McLaren and D. Tabor, "Friction of Polymers at Engineering Speeds: Influence of Speed, Temperature, and Lubricants," *Proceedings of the Lubrication and Wear Convention 1963*, Institution of Mechanical Engineers, London, 1964, pp. 210–215.

32. W. T. Thomson, *Theory of Vibrations with Applications*, 2nd ed., Prentice-Hall, Englewood Cliffs, NJ, 1981, pp. 48–52.

33. R. D. Frantz, "An Experimental Study of Fretting Corrosion at a Bearing/Cartridge Interface," M. S. thesis, Virginia Polytechnic Institute and State University, Blacksburg, VA, 1983, pp. 58–59.

34. M. B. Peterson, "Design Considerations for Effective Wear Control," in M. B. Peterson, and W. O. Winer, Eds., *Wear Control Handbook*, American Society of Mechanical Engineers, New York, 1980, pp. 451–457.

35. J. D. Jones and N. S. Eiss Jr., "Effects of Chemical Structure on the Friction and Wear of Polyimide Thin Films," in L-H. Lee, Ed., *Polymer Wear and Its Control*, ACS Symposium Series 287, American Chemical Society, Washington, DC, 1985, pp. 135–148.

36. N. S. Eiss Jr., J. H. Warren, and S. D. Doolittle, *Wear*, **38**, 125 (1976).

37. D. H. Buckley, *J. Vac. Sci. Technol.*, **13**, 88 (1976).

38. H. Czichos, *Tribology*, Elsevier, Amsterdam, 1978, p. 267.

39. F. Hendricks, "Squeeze Bearing Levitated Sliders for Magnetic Storage," in B. Bhushan and N. S. Eiss Jr., Eds., *Tribology and Mechanics of Magnetic Storage Systems*, Vol. IV, Special Publications SP-22, Society of Tribologists and Lubrication Engineers (formerly American Society of Lubrication Engineers), Park Ridge, IL, 1987, pp. 26–35.

40. R. B. Lewis, *Mech. Eng.*, **86**, 32 (1964).

41. D. E. Newland, *An Introduction to Random Vibrations and Spectral Analysis*, Longman, London, 1975, pp. 150–166.

42. Ch. Colombie, Y. Berthier, A. Floquet, L. Vincent, and M. Godet, *ASME J Tribology*, **106**, 185 (1984).

43. Y. Berthier, Ch. Colombie, L. Vincent, and M. Godet, *J. Tribology*, **110**, 517 (1988).

44. S. C. Gordelier and J. Skinner, "Graphite Wear Against Rough Surfaces—The Effect of Sliding Distance and Interface Gas Flow," in D. Dowson, M. Godet, and C. M. Taylor, *Wear of Nonmetallic Materials*, Mechanical Engineering Publication, Edmunds, England, 1978, pp. 202–209.

45. M. Chitsaz, "The Effects of Rubber Modification on Friction and Wear of Epoxy Networks," Doctoral dissertation, Virginia Polytechnic Institute and State University, Blacksburg, VA, 1987.

46. J. E. Freund, *Modern Elementary Statistics*, 4th ed., Prentice-Hall, Englewood Cliffs, NJ, 1973, pp. 249–256.

47. J. E. Freund, *Modern Elementary Statistics*, 4th ed., Prentice-Hall, Englewood Cliffs, NJ, 1973, pp. 389–416.

48. M. G. Natrella, *Experimental Statistics*, NBS (NIST) Handbook 91, National Institute of Standards and Technology, Washington, DC, 1963, pp. 3.22–3.30.

49. R. A. L. Rorrer, H. H. Mabie, N. S. Eiss Jr., and M. J. Furey, *Tribology Trans.*, **31**, 98 (1988).

50. M. G. Natrella, *Experimental Statistics*, NBS (NIST) Handbook 91, National Institute of Standards and Technology, Washington DC, 1963, pp. 12.1–12.21.

51. W. G. Cochran and G. M. Cox, *Experimental Designs*, 2nd ed., Wiley, New York, 1957, pp. 244–292.

52. W. G. Cochran and G. M. Cox, *Experimental Designs*, 2nd ed., Wiley, New York, 1957, pp. 6–9.

53. D. E. Hinkle, W. Wiersma, and S. G. Jurs, *Applied Statistics for the Behavioral Sciences*, Rand McNally, Chicago, IL, 1979, pp. 150–235.

54. P. J. Blau, *Friction and Wear Transitions of Materials: Break-in, Run-in, Wear-in*, Noyes Publications, Park Ridge, NJ, 1989, pp. 33 and 189–190.

Chapter **4**

MECHANICAL CHARACTERIZATION OF COMPOSITE MATERIALS

Dale W. Wilson and Leif A. Carlsson

1   INTRODUCTION

This chapter introduces the methods used for mechanical property characterization of fiber reinforced polymer composite materials. It covers methods for the determination of fundamental elastic and strength properties, but it does not attempt to investigate fatigue or component level testing issues. Composite materials exhibit complex elastic and failure behavior, making them more difficult to characterize than traditional isotropic, homogeneous materials. With composites it is important to understand the test method and material characteristics in order to interpret results properly.

The test methods that apply to the characterization of composite materials are addressed by the activities of several agencies including the American Society for Testing and Materials (ASTM), Military Handbook 17 (MIL-HDBK-17), Society of Automotive Engineers (SAE), National Aeronautics and Space Administration (NASA), and Suppliers of Advanced Composite Materials Association (SACMA). Of these ASTM has written the most widely accepted standards, which form the basis for most composite test methods.

The mechanical properties obtained from a mechanical characterization are controlled by several material parameters that are in turn influenced by how the material is processed and by its composition. In Section 1 we summarize the material structure property relationships and basic mechanical behavior. The objective is to provide enough background to interpret results from mechanical tests properly and, where necessary, to make judgments about the type of test to use.

While composites use the typical array of mechanical test instrumentation for imparting loads and measuring strain and displacements, certain characteristics of composite systems dictate the selection of certain systems and specific instrumentation for the best possible results. In Section 2 we describe typical test instrumentation and discuss special requirements for composite testing.

In Section 3 we describe specific mechanical characterization methods for fiber reinforced composite materials at three levels: constituent fibers and matrix polymers, composite laminas, and composite laminates. The measurement of tension, compression, shear, flexure, and fracture properties will be covered. When there is more than one suitable test method for determining a property, the characteristics of each will be presented on a comparative basis.

1.1 Types of Composite Material Systems

A fiber reinforced plastic composite consists of a matrix phase, which is a thermosetting or thermoplastic polymer, reinforced by one or more fiber systems. The typical microstructure of a composite is shown in Figure 4.1. Common fiber reinforcements include glass, carbon, boron, aramid, and silicon carbide. The fiber reinforcement can be either continuous or discontinuous. If continuous, the fiber reinforcement can be unidirectional or in multidirectional woven textile form. Depending on the purpose of the reinforcement, discontinuous reinforcements can vary in length from fractions of a millimeter to tens of millimeters in length.

The properties of the matrix and the fiber, the relative composition of the constituents, and the orientation of the fiber reinforcement control the fundamental properties of a composite. The exact property relationships are governed by the micromechanics of the system. The simplest system is the continuous fiber unidirectional composite. In this system the reinforcing fibers are straight, oriented in well-defined directions, distributed uniformly throughout the material, and occupying a known volume fraction of the system. Fabrics are more complex in that weaving introduces an undulation or crimp in the fiber and the dispersion of the fiber is less uniform. Discontinuous fiber systems are the most complex. In discontinuous fiber systems the orientation of the fibers is statistically distributed and unknown, the length of reinforcing fibers is not uniform, and the uniformity in dispersion of the reinforcements is unknown. The micromechanics of such a system can only be defined through statistical procedures. References [1-4] give detailed information on micromechanical models describing composite behavior.

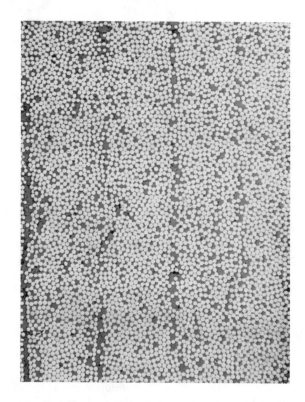

Figure 4.1 Typical microstructure of a composite (G30-500 carbon fiber epoxy).

From micromechanics it is clear that the first level of characterization for composites is to measure the properties of the constituent phases. The measured properties for matrix and fiber materials can then be used to determine the expected properties of composite systems. Of course the fiber-matrix interface also plays an important role in the translation of constituent properties into composite properties and it is very difficult to characterize. The matrix phase is characterized using the methods generally used for any other polymer system and they are not covered in detail here. Advanced composite fibers do require special characterization methods and they are discussed in this chapter.

Continuous fiber composites, like carbon/epoxy, are typically laminated to form structural materials as shown in Figure 4.2. Each single ply, or lamina, has orthotropic material properties that are typically described in relation to a material coordinate system as shown in Figure 4.3. The material is strong in the fiber direction and very weak transverse to the fibers. The ply or lamina is the most fundamental level of composite property for a continuous fiber composite. Using lamination concepts, laminates are formed by orienting the plies at angles to an arbitrary laminate coordinate system (Figure 4.3) in a way that overcomes

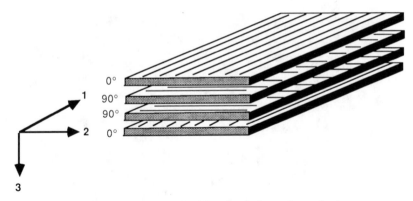

Figure 4.2   Lamination of unidirectional plies to form a laminate.

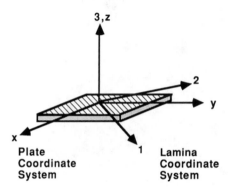

Figure 4.3   Definition of coordinate systems for composite laminates. The plate coordinates are labeled x, y, and z; and the lamina coordinates are labeled 1, 2, and 3.

the weak transverse properties. The orthotropic properties of each ply, as well as the ply orientation, determine the effective properties of the laminate. These constitute two important levels of characterization for continuous fiber composites—the lamina and laminate properties.

Normally, discontinuous fiber composites are molded into functional parts and have a microstructure as shown in Figure 4.4. Unlike the composites for continuous fiber systems, there is no fundamental unit, such as a ply, for composite characterization. The characterization of discontinuous fiber composites must be handled very carefully. As mentioned earlier, the properties are determined by the volume fractions of constituent phases, often including a particulate filler, the length-to-diameter aspect ratio of the reinforcement, the uniformity on dispersion of the reinforcement, and the fiber orientation distribution of the fiber reinforcement. This is strongly influenced by the way the material is processed. The characterization of discontinuous fiber composites is similar to lamina characterization except a reference coordinate system must be defined before the orthotropic material properties can be measured in relation

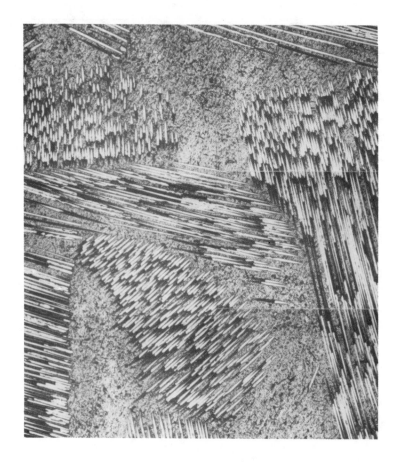

Figure 4.4   Microstructure of a discontinuous fiber composite.

to the chosen coordinate system. The material is treated as an orthotropic sheet. Occasionally the material develops a layered microstructure that is very similar to a laminated structure and is then handled like a laminate.

1.2   Mechanical Properties of a Composite System

The mechanical properties of a composite material are the elastic and strength properties of the material. For most practical composite material systems the elastic properties are defined by an orthotropic set of nine independent elastic constants. An orthotropic material has three orthogonal planes of symmetry. A sheet molding compound is usually an example of an orthotropic material. Many materials are transversely isotropic; one plane is isotropic, which reduces the number of independent constants to five compared with a typical isotropic material that has only two independent elastic properties. Many typical unidirectional continuous fiber composites are transversely isotropic. Designers use a fourth-order tensor equation to describe the elastic behavior of composite systems, but the constants can be described in simpler engineering terms.

That there are between five and nine independent elastic constants oversimplifies the reality of testing for mechanical properties. There are other properties, such as fracture toughness and flexural strength and stiffness, that are also useful in characterizing the performance of a composite material. Composites behave differently in compression than in tension; the elastic and strength properties must be characterized in both tension and compression loading. Fracture toughness can be measured in mode I, a crack opening mode; mode II, an interlaminar shearing mode; and mode III, an interlaminar scissoring mode.

The basic concept of homogeneity governs the definition of material behavior for traditional metallic material systems. The homogeneity assumption holds that the microstructural features of the material are small enough to be inconsequential to the average behavior of the material on a macroscale. Unfortunately this is not always true for composites, especially when strength and fracture are considered. Fabrics and laminates are very inhomogeneous in character. The scale of the homogeneity of the composite system affect the type of instrumentation used to measure deformation and preclude the use of the continuum mechanics theory to determine properties. Laminate theory and special approximations are used to analyze results of composite systems that do not allow the use of homogeneous theory. Fortunately, in some simple cases continuous fiber lamina do allow the use of homogeneous elastic theory to define strength and modulus.

The best way to describe the engineering properties for a lamina is to look at the lamina coordinate system as shown in Figure 4.5 and consider tension, compression, and shear loadings. The properties, which number 27, then become those described in Table 4.1. If the material is transversely isotropic, then the starred properties do not need to be determined. The properties in Table 4.1 are the strength and modulus properties. While not elastic properties, the fracture properties of materials are also often measured to compare the toughness of material systems. These properties are G_{Ic} the critical mode I strain energy release rate, and G_{IIc}, the critical mode II strain energy release rate. Flexural properties are also determined routinely. The flexural properties result from bending the material to produce tension, compression, and shear stresses in the material. The result is more a structural property than an intrinsic material property, but it is very useful in materials screening and quality control.

Laminate properties that are determined for composite systems are identical to those for the foregoing lamina except a laminate coordinate system is employed. The subscripts on the properties defined in Table 4.1 are changed to x, y, and z and the properties become the effective laminate properties. The word effective is very important because it signifies that the measured response is an average response through the material thickness. In reality the stress in each

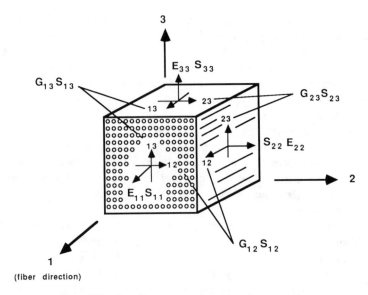

Figure 4.5 Coordinate system for composite properties.

ply of a laminate is different; it is the product of its stiffness and the assumed uniform deformation of the test specimen.

1.3   Test Equipment and Fixturing

The mechanical characterization of composite materials requires special test equipment for specimen load introduction and deformation measurement and fixturing to interface the specimen to the test machine while the proper state of stress is generated in the specimen. Standard equipment is available for load introduction and strain measurement, but test fixturing is specially designed for many of the composite tests. The drawings and specifications for standard composite test fixtures are available in the test standards, but many fixtures are not available commercially and it is often necessary to have a fixture machined at a local machine shop.

1.3.1   Test Machines

Loading is usually applied by a universal test machine similar to the one shown in Figure 4.6. The essential characteristics of this test machine are a very stiff base, a movable crosshead, a stationary crosshead mounted at the top of the test frame, and a load measuring device. The movable crosshead is driven precisely by twin screws at specified rates relative to the stationary member to generate either tension or compression loads. A load cell is mounted in either of the crossheads or on the base as necessary for the test setup desired. Most machines provide a chart recorded to plot a record of load and crosshead displacement

Table 4.1 Matrix of Properties Necessary to Characterize a Composite Material

Symbol	Property
E_1^t	Tensile modulus in the fiber direction
E_2^t	Tensile modulus transverse to the fiber
E_3^t	Tensile modulus transverse through the thickness[a]
E_1^c	Compression modulus in the fiber direction
E_2^c	Compression modulus transverse to the fiber
E_3^c	Compression modulus transverse through the thickness[a]
G_{12}	Shear modulus in the 1-2 plane
G_{13}	Shear modulus in the 1-3 plane
G_{23}	Shear modulus in the 2-3 plane[a]
X_1^t	Tensile strength in the fiber direction
X_2^t	Tensile strength transverse to the fiber
X_3^t	Tensile strength through the thickness[a]
X_1^c	Compression strength in the fiber direction
X_2^c	Compression strength transverse to the fiber
X_3^c	Compression strength through the thickness[a]
S_{12}	Shear strength in the 1-2 plane
S_{13}	Shear strength in the 1-3 plane
S_{23}	Shear strength in the 2-3 plane[a]
v_{12}	Major Poisson's ratio in the 1-2 plane
v_{13}	Major Poisson's ratio in the 1-3 plane
v_{23}	Major Poisson's ratio in the 2-3 plane[a]
ε_1^t	Ultimate tensile strain in the fiber direction
ε_2^t	Ultimate tensile strain transverse to the fiber
ε_3^t	Ultimate tensile strain through the thickness[a]
ε_1^c	Ultimate compression strain in the fiber direction
ε_2^c	Ultimate compression strain transverse to the fiber
ε_3^c	Ultimate compression strain through the thickness[a]

[a]These properties do not need to be determined when the material is transversely isotropic.

during a test. This type of universal test machine operates in a displacement controlled mode only, using displacement to generate stress in the attached test specimen.

There are also servohydraulic test machines for load application. Servohydraulic machines are constructed with a base, a vertical test frame, and a stationary crosshead (see Figure 4.7). Loading is achieved through a hydraulic actuator mounted either in the base or in the stationary crosshead. One principal difference with the servohydraulic machine is that precise load, stroke, or strain control testing is possible. Also, programmable load cycling is possible for fatigue testing. Many modern machines are controlled by computers that can generate complex waveshapes and frequency spectra to simulate actual fatigue

Figure 4.6   Picture of typical universal testing machine with electromechanical screw drive mechanism.

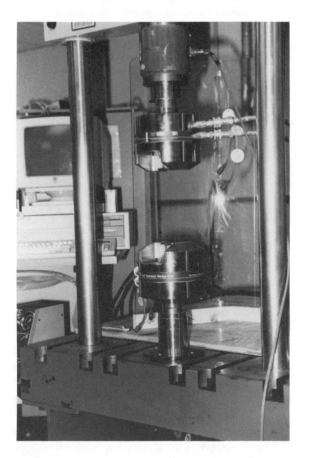

Figure 4.7   Picture of typical servohydraulic testing machine.

conditions. In general the capabilities of this type of test machine are not required for simple characterization of static properties of composite systems.

To test composite materials a universal test machine with greater than or equal to 100-kn capacity is recommended because of its excellent stiffness characteristics and load capacity; longitudinal tension and compression properties of some composite specimens require this capacity. On the other hand, fiber tests, transverse tension, and flexural tests require much lower load sensitivity. Universal test machines allow interchangeability of load cells to accommodate different testing requirements, and when needed small capacity load cells can be used in the 100-kn frame.

1.3.2   Strain Measurement

There are many ways to measure displacement or deformation of materials. The three most common transducers used for this purpose are strain gauges†, extensometers, and linear variable differential transformers (LVDT). The purpose of each transducer is to convert a physical deflection or change in dimension of the test specimen to a calibrated electrical signal representing displacement or strain. Special considerations are necessary when using strain transducers on composite materials and they are discussed.

A typical strain gauge is shown in Figure 4.8. The area of the active portion of the grid and the resistance of the grid are important. The grid area must be large enough to average deformation over a representative area of the specimen. If the material is not very homogeneous, such as woven fabrics, the

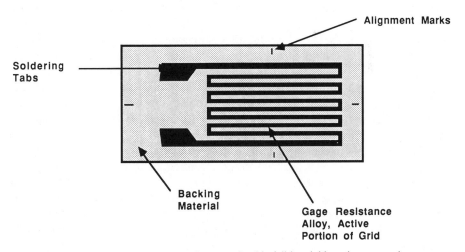

Figure 4.8   Illustration showing key features of a thin foil bondable resistance strain gauge.

†The word *gauge* is usually spelled *gage* in the industry; however, here, for indexing purposes, we use the conventional form.

grid must be large enough to cover at least one basic unit cell of the structure. The resistance of the grid used on a composite is important because of gauge heating effects. A constant voltage is applied to the gauge. The strain gauge is normally located in a balanced bridge circuit at zero strain. When the specimen is deformed the strain gauge resistance changes proportional to the deformation, which produces a calibrated voltage offset representing a precise level of strain. Because the power dissipated by the strain gauge is a function of the voltage squared divided by the resistance of the gauge, higher resistance gauges cause less heating for a given voltage. Gauge heating is a problem with composites because the polymer matrix usually does not conduct heat very well, which allows heat to build up in the gauge. The resulting temperature change causes a resistance change, which is falsely recorded as an apparent strain. Therefore, the use of 350-Ω strain gauges is recommended for composite testing, and the excitation voltage should be less than 5 V.

Tests where a temperature transient is allowed also produce an apparent strain that is the result of the change in gauge resistance and the thermal coefficient of expansion mismatch between the gauge and the test material. This will not occur if the test is conducted at an elevated temperature but held constant. However, the appropriate calibration factor (gauge factor) should be employed since the gauge factor is a function of temperature.

The effectiveness of strain measurement using bonded foil strain gauges depends on a high quality bond between the gauge and the test specimen. The specimen surface should be prepared exactly as instructed in the adhesive literature on gauge application. The adhesive must have temperature capability over the temperature range of the test. Failure to properly bond the gauge produces variable and erroneous results. Bondable foil strain gauges cannot be reused after a test.

Extensometers use a full bridge of strain gauges mounted on a specially designed spring beam. Special arms are attached to the calibrated, instrumented beam with a separation equal to the gauge length of the extensometer. The advantage of extensometers is that the arms are clipped to the specimen; this allows quick installation and makes the unit reusable. The gauge length of an extensometer should be chosen to match the homogeneity of the specimen and to provide the strain resolution necessary for the material.

Extensometers usually have knife edges that contact the specimen to prevent slippage. The knife edges are held to the specimen using spring-loaded clips or rubber bands. Certain specimens can be damaged by these knife edges, and a measured low apparent strength results. Controls should always be run to ensure the extensometer is not affecting the strength of the material being tested. If strength is adversely affected, separate strength and modulus specimens may be necessary.

Linear variable differential transformers, like extensometers, are mechanically attached to specimens and hence have similar operational characteristics. Linear variable differential transformers are more bulky and they are more often used for direct measurement of displacements, such as the center span displacement of a flex beam or the displacement of the movable crosshead of the test machine relative to the fixed crosshead.

1.3.3 Fixturing Issues

Fixturing is critically important to composite testing. Each test requires its own special fixture that works in combination with its mating test specimen. The special characteristics of each fixture are described as part of the individual test procedures. The general function of the test fixture is twofold: (1) to transfer loads or displacements from the test machine to the test specimen, and (2) to effect the transfer of load or displacement such that it produces the correct state of stress in the specimen test section.

No fixture functions perfectly, especially in generating a perfect state of stress in the test specimen. Fixtures must provide for reproducible alignment of the specimen in the test machine. Specimens should be easy to insert and to remove after testing. The fixture must be strong and stiff enough not to change the characteristics of the state of stress in the specimen during the test. It must also be constructed of hardened materials that will not wear excessively with repeated use. In addition the fixturing must function in whatever environment the test is performed. If a fixture is too bulky, high-temperature testing is very difficult.

All fixtures must be inspected routinely before testing to ensure that they are not worn or damaged in any way that will affect the test results. With use all fixtures wear and eventually produce increased variability in test results.

1.3.4 Environmental Conditioning

Environmental conditioning is perhaps one of the most controversial aspects of composites testing. With composites, it is not just the temperature at which the material is to be tested that is important. The whole temperature and moisture history of the specimen influences the properties.

Specimens should be preconditioned before testing by exposure to the specified temperature and moisture environment for the specified duration. This can be accomplished by using conventional environmental chambers or temperature-controlled baths. Moisture conditioning is often specified in one of three ways: exposure at specific temperature and moisture conditions to attain a target percentage weight gain, exposure for a specified time duration at defined temperature and moisture conditions, or exposure to attain equilibrium moisture gain at specific temperature and moisture conditions. Since diffusion constants for composite systems are often very small, sometimes accelerated conditioning practices are employed. For example, specimens can be soaked in water for two weeks to attain a certain moisture content instead of being conditioned for months at 95% relative humidity to attain the same moisture content.

Accelerated conditioning is usually not equivalent to normal conditioning.

The aging of the material is a rate process that depends on path; thus, a simple time–temperature superposition principle does not always hold.

Any composite system to be conditioned and tested under specified environmental conditions should be fully characterized using infrared (IR) spectroscopy, thermomechanical analysis (TMA), and differential scanning calorimetry (DSC) to characterize the state of cure, moisture composition, and glass transition temperature (T_g) of the material. Once known, this information defines the preconditioned state of the material. The material can then be subjected to elevated temperature or combined elevated temperature and moisture conditions for either the specified duration or until a defined equilibrium state is achieved. Often a moisture content is defined by a simple target percent weight increase of the specimen caused by moisture uptake. Once conditioned, specimens should then again be characterized by analytical equipment. This is best done on travelers, which are dummy specimens conditioned along with the test specimens for this purpose. The test specimens cannot be used since the analytical tests will influence the moisture content of the specimen and render it useless for testing.

1.4   Fabrication of Test Specimens

The fabrication of test specimens is one of the most critical aspects of composite testing. As mentioned earlier, a composite material is not formed until a panel or part has been processed. The volume fraction of fiber to matrix, the void content, and the uniformity of fiber wetout in the part are controlled by the processing of the test panel. It is very important to control the processing of the test panels properly so that the resulting composite test panel is representative of the typical composite that the material will produce.

1.4.1   Specimen Lay-Up

Every test specimen has a defined laminate configuration. The prepreg material must be cut to the proper size and laid up in the proper orientation. When sizing test panels, about 25–50 mm of scrap should be included around the border to allow trimming after cure. During the lay-up a reference edge must be established and each ply oriented accurately with respect to the reference border. Before cure, the laminate reference edge is accurately scribed with a reference line that is used to maintain alignment when the cured plate is trimmed.

1.4.2   Specimen Machining

We normally machine test specimens out of panels by using a diamond wafing saw. A cut is first made along the reference edge and then all subsequent cuts are made relative to the reference edge to preserve the accuracy of the fiber orientation in the panel. The diamond cutoff wheel produces a very smooth surface along the cut edges, and no further finishing is usually required.

In specimens requiring holes a diamond core bit provides the best quality of hole. Also, several special drill bits are available; and when they are used with templates to guard against punch-through delamination, satisfactory results are produced.

Normally the precision of the diamond wafing saw is ± 0.05 mm, which is accurate enough for composite specimens. In certain cases very high dimensional precision is desired; for instance, in testing end-loaded compression samples, a surface grinder can be used to grind specimens to the required precision.

If a specimen is designed to have tabs, the tabs are bonded into place before the specimen is machined to its final shape. The tab material is typically 3.2-mm-thick [0/90] glass/epoxy or a woven fabric glass/epoxy material. A bevel is machined onto the tab and then the tab is bonded to a plate of the test panel by using special bonding jigs to assure alignment of the tabs with the specimen reference edge. Individual specimens are cut from the tabbed panel again in alignment with the reference edge.

Good quality composite specimens should be of uniform dimensions, have a precise fiber alignment, and possess high quality finish on machined edges. There should be no evidence of delamination along machined edges. The laminate should be of high quality, with no dry fiber regions, voids, or other obvious flaws. If available, tools such as ultrasonic c-scan should be employed to evaluate composite specimens nondestructively for flaws prior to fabrication and testing. Flawed specimens should be discarded.

2   CHARACTERIZATION OF CONSTITUENT PROPERTIES

Primarily, the first level of characterization for composite systems serves the function of quality assessment—quality of the fiber and resin components and quality of the fundamental composite system. Vendors purchasing fibers or resins want to know the physical and mechanical properties of their lots. Most companies have materials specifications for acceptance that are based on properties demonstrated for the product during a materials qualification test program.

Two important properties of cured composite laminates are void content (voids are small holes in the resin phase) and fiber volume fraction. High void content causes poor properties in a composite laminate, because voids are small flaws. Volume fraction influences the strength and stiffness of the composite system as defined by the equations of micromechanics. Laminates containing significantly different volume fractions of fiber exhibit different mechanical properties and cannot be compared directly without some type of normalization. To determine void content or fiber volume fraction, the physical property of density must be determined for the fiber, the resin, and the composite.

The material in this section covers the determination of density, void content, and fiber volume fraction for composite systems and also the mechanical characterization of polymerized matrix systems and fibers. Where possible references to ASTM standards are given.

2.1   Density Measurement

Methods that are commonly used to measure density of composite materials and their components are the Archimedes, the sink–float, and the density-gradient methods. The first two methods are covered in ASTM D3800 [5], while the third is covered by ASTM D1505 [6].

The Archimedes method compares the weight of a material sample in air and while it is immersed in a lower density liquid with known density. The buoyancy force on the immersed sample is simply the difference in the weight obtained in the two media. The sample volume is obtained by dividing the buoyancy force by the density of the liquid. The density of the sample is simply the initial weight in air divided by the sample volume.

When performing the Archimedes test, ensure that the sample is fully wetted by the liquid, with no entrapped air bubbles, since air bubbles will affect the buoyancy force measurement. Similarly, the method makes the tacit assumption that the sample is voidfree. The presence of voids leads to error in the density determination.

The sink–float and the density-gradient methods employ similar principles. We start the sink–float method by immersing a sample in a liquid of lower density and then slowly adding a higher density liquid until the sample floats to the top. The density of the resulting mixture of liquids is then measured using either a hydrometer or pycnometer. The density of the sample is equal to the density of the liquid mixture in which it just floats.

The density-gradient method employs a density-gradient column, which is a column containing a liquid mixture of linearly varying density from bottom to top. The column is kept at a precise temperature and is calibrated by inserting calibration floats of precisely known density. The sample to be measured is then dropped into the density-gradient column and allowed to settle to an equilibrium position. The density is determined by interpolating between the sample position and the positions of the two nearest calibration floats.

When testing the density of fibers, sufficient length of fiber should be used to weigh at least 0.15 g, nominally 3 m or greater. Neat resin and composite samples are easier to handle and also should be of appropriate size and volume to insure an accurate measurement.

2.2   Fiber Volume Fraction

The fiber volume fraction of polymer composites is usually determined by either the ASTM D2584 [7] or the ASTM D3171 method [8]. The ASTM D2584 method is used for glass fibers and involves burning off the matrix, while the ASTM D3171 method is for carbon and aramid fiber composites and uses a chemical digestion of the polymer matrix phase.

The two methods are needed because glass fibers are very resistant to oxidation at the temperatures required to burn off polymer matrix materials while being susceptible to attack by the acids employed in matrix digestion. Conversely, aramid and carbon fibers are readily oxidized by the burn-off technique and are more stable in the acids used to dissolve the matrix polymer. It is important that the fiber is not affected by the matrix removal process.

Aside from the method employed to remove the matrix phase, the methods are similar. The density of the fiber must be known and the density of the composite must be determined before the matrix is removed when the fiber content by volume is to be determined. The dry weight of the composite specimen is determined to the nearest 0.0001 g. The matrix is then removed using the technique appropriate for the fiber and polymer systems comprising the composite. For glass fibers the burn-off technique should be used. For carbon or aramid fibers three different procedures are available.

1. Hot nitric acid—used for epoxy resin matrix materials.
2. Hot sulfuric acid followed by hydrogen peroxide—used for phenolics, polyimides, and poly(ether ether)ketone (PEEK).
3. Ethylene glycol–potassium hydroxide solution—used for anhydride cured epoxies.

With these procedures care must be taken not to overdigest the material since fiber loss can occur with extended exposure to the caustic chemicals used for matrix digestion. Once the matrix is removed the fiber is dried and weighed. Using the density and weight information we can determine the fiber weight percent from (1).

$$\text{fiber weight\%} = \text{fiber weight/composite weight} \times 100 \qquad (1)$$

or we can find the fiber volume fraction from (2).

$$\text{fiber volume fraction} = [(W_f/\rho_f)/(W_c/\rho_c)] \qquad (2)$$

where W_f is the fiber weight, ρ_f is the fiber density, W_c is the composite weight, and ρ_c is the composite density.

Potential sources of error in this procedure are caused by the loss of fiber because of overdigestion. To correct for this, fiber weight loss can be measured as a function of time and used to formulate a correction factor. By knowing how long the fiber is immersed in the chemicals, we can use the correction factor to estimate fiber weight loss.

2.3 Void Content

Void content determination is very similar to fiber volume fraction determination. The procedure for void measurement is given in ASTM D2734 [9]. It utilizes techniques employed in both the density and fiber volume fraction methods.

The basic principle involves determining the voidfree, theoretical density and the real density of the material. The void content is proportional to the difference between the theoretical and the measured density of the composite. For a good composite the void content should be less than 1%. Determination of the theoretical density for the composite involves measuring the density of the fiber and the density of the resin, measuring the fiber weight fraction, and then calculating the composite density using

$$\rho_c = 1/(w_r/\rho_r + w_f/\rho_f) \tag{3}$$

where ρ_c is the density of the composite, w_r is the weight of the resin, w_f is the weight of the fiber, ρ_r is the density of the resin, and ρ_f is the density of the fiber.

The fiber weight fraction is found by weighing the sample initially, dissolving off the matrix, and weighing the remaining fiber. The difference between the total weight and the fiber weight is the resin weight. The void content is found as a percentage using the following equation:

$$\text{void content} = 100(\rho_{ct} - \rho_{cm})/\rho_{ct} \tag{4}$$

where ρ stands for density and the ct and the cm subscripts refer to the theoretical and measured densities of the composite, respectively.

2.4   Matrix Test Methods

Matrix mechanical properties are rarely determined for quality control purposes. They are mostly determined during materials development to evaluate performance to define basic properties for product data sheets. Resin developers use mechanical properties to rank new candidate materials before moving into composite characterization, thereby reducing the number of systems that must be evaluated in composite systems. We determine these properties for use in micromechanics models that predict composite properties from the constituent properties or for research and development of new matrix systems. In materials development, often only small quantities of material are available to evaluate performance. Carrying out mechanical evaluation of neat resin samples can help us to screen the best systems for further development. The composite characterization is needed as the final step since the influence of the interface is not accounted for in neat resin properties. Chemical characterization is the chief quality control tool, but it is not covered here.

The testing of matrix properties is not covered in detail in this section because the methods employed closely parallel those used for laminates. The mechanical properties that are normally measured for the matrix material are tensile strength, tensile modulus, shear strength, shear modulus, and coefficient of thermal expansion.

2.4.1   Specimens

In most cases neat resin materials are cast into bars for testing. The size of the bars depends on material availability and on the characteristics of the resin.

Some materials can be cast into voidfree thick bars or dogbone specimens for testing. Other materials are best cast into thin films or sheets. To the extent possible specimen design should follow the guidelines specified for the test procedure used.

2.4.2 Testing

We can best characterize matrix materials by using the same procedures used for lamina evaluation. The standard tension test based on ASTM D3039 [10] or ASTM D638 [11] can be used for tension properties when thick dogbone samples can be fabricated. Shear is best measured by using the V-notched beam method [12-15]. The details of the V-notch beam shear procedure are outlined in Section 3.4. The chief deviation from the procedures recommended for composites is the reduction in crosshead speed for matrix materials. Since the materials are usually very brittle, exercise extreme care in handling and during the test execution.

Strain measurement should be accomplished through the use of an extensometer or strain gauges. If an extensometer is used, take care to monitor the effect of the sharp knife edges of the extensometer on strength. If strain gauges are used, 350-Ω resistance gauges should be used, not 120-Ω gauges. Excitation voltages should be kept as low as possible to avoid gauge heating. The polymer cannot dissipate heat quickly and the heat buildup can cause temperature-induced resistance change causing error in the strain measurement. The use of pulsed excitation instrumentation also avoids this problem.

2.4.3 Issues

The presence of voids in castings may influence strength and modulus determination. The quality of specimens is important. In addition always remember that neat resin properties are not necessarily representative of those that will be achieved by the resin in a composite. The interface and fiber packing geometry play a strong role in property translation.

2.5 Fiber Testing

The purpose of characterizing constituent fiber and matrix properties depends on who is performing the characterization. Material producers measure properties for research and development of new products and certification of commercial products. End users test materials for quality control before acceptance of purchased materials.

Fiber tests are used extensively for fiber quality control and product certification. They are used mostly by fiber producers for lot certification and by prepreggers for quality control of incoming material before production to assure that fibers meet the required specifications. Fiber test methods are precisely defined and controlled for this reason. Fiber producers do use fiber characterization in the research and development of new fibers. The results from fiber tests allow quick evaluation of how processing conditions are affecting fiber properties.

Fiber characterization methods are designed to measure fiber strength and modulus. There are two types of tests used: a single filament test [15] and an impregnated strand test [16].

2.5.1   Single Filament Tests

The single filament test can be used to determine fiber strength and modulus, but it is not very accurate for modulus determination. Instrumentation cannot be applied and crosshead motion must be used for strain determination if modulus is desired. The test is used primarily for brittle, high modulus fibers or in situations where actual fiber data are desired for process development purposes.

In the test a single filament is separated from a dry tow of fibers and mounted on a cardboard tab as shown in Figure 4.9. The tabbed specimen is then inserted in a test machine, the supporting tab carefully burned away, and the unsupported single filament tested to failure. The test requires a 500-g capacity load cell. To obtain meaningful data, a statistically significant number of samples must be tested. Because of high variability in fiber strength, large numbers of specimens are usually necessary. This test is only used for very special purposes; usually the impregnated strand test is preferred.

One source of variability in the test data is the fiber yield used to determine fiber cross-sectional area. Yield, the weight per unit length, is determined on the dry tow from which the fibers are selected prior to separating the individual filaments for specimen mounting. This means that an average denier is used, not specific data for each specimen. Care must be exercised to assure that coalesced fibers are not chosen. Normally, each strand specimen should be inspected for quality under an optical microscope prior to testing.

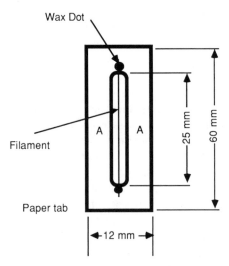

Figure 4.9   Single filament test specimen.

2.5.2   Impregnated Strand Tests

The impregnated strand test is used to measure strength and modulus also, but it yields inherently more consistent results than the single filament method. In this test a dry fiber tow consisting of thousands of individual filaments is soaked in a polymer resin and cured. The resulting strand is then tested in tension to failure. An extensometer can be mounted to measure strain accurately. By coating the fibers with resin, load is transferred to the fibers more uniformly and strength is averaged over thousands of filaments, giving more consistent results.

In strand testing specimen preparation is critical; too little resin leaves sections of fiber dry and results in poor strength translation. Most of the time the fiber is run through a resin bath and pulled through a nozzle to develop a consistent resin content in the tow. The resin-impregnated tow is then wrapped on a rack under slight tension and allowed to dry. The dimensions of the rack are such that the final strand has a length of 330 mm. Tabbing is optional. Most tabs are now fabricated by using a room temperature adhesive and heavy paper (freezer paper) as shown in Figure 4.10.

Prior to fabrication the yield, weight per unit length, of the fiber tow must be determined. The yield is determined by measuring the weight to the nearest 0.0001 g of a 90-cm length of the strand. The density must also be determined for

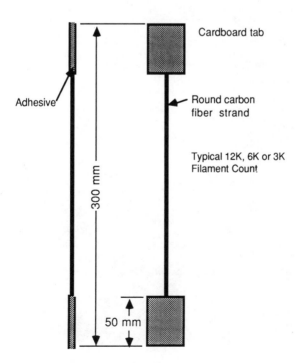

Figure 4.10   Impregnated strand test specimen.

the fiber. The yield divided by the density gives the effective fiber area

$$A_f = W/\rho_f \qquad (5)$$

where A_f is the effective fiber cross-sectional area, W is the yield in grams per meter, and ρ_f is the fiber density in grams per cubic meter.

The test is usually conducted in a universal testing machine with pneumatic rubber-faced grips as shown in Figure 4.11. Sometimes emery cloth is employed on the grip faces to facilitate gripping, especially when no tabs are used. A 5-cm gauge length, 10% elongation extensometer is employed to measure strain in the fiber during loading, and the load versus strain chart is recorded during testing. The strength is determined from the equation

$$\sigma^{\text{ult}} = P/A_f \qquad (6)$$

Figure 4.11   Strand test setup showing pneumatic grips.

Where A_f is the effective fiber area, P is the ultimate load carried by the fiber, and σ^{ult} is the ultimate fiber strength. The effect of the resin is ignored in most cases for aerospace grade high strength, high modulus fibers. The modulus is determined from the slope of the stress–strain curve. Not all fibers exhibit a linear elastic strain behavior to failure. Care must be taken to determine modulus consistently during fiber testing, for the value of modulus determined can change depending on how the calculation is performed. There are many choices for modulus determination, such as the tangent modulus at 0.5% strain, secant modulus between 1000 and 6000 microstrain, and initial tangent modulus. Once chosen, the method for determination of modulus must be adhered to consistently and of course be defined clearly. Probably one of the most sensible methods, which is consistent with many tests being written by SACMA, is the use of modulus measured between 1000 and 6000 microstrain.

2.6   Summary of Constituent Characterization

Methods have been presented for characterizing the matrix properties, fiber properties, density, void content, and fiber volume fraction. The constituent property information can be used in micromechanics models to determine estimates of composite properties or to test the quality of incoming material before prepregging. Void content and fiber volume fraction should be determined on all laminates. Fiber volume fraction is critical; when comparing properties, the volume fraction must be held constant. Since this is very hard to do, data are often corrected to a specific reference volume fraction by assuming a linear relationship between fiber volume fraction and the property, modulus, or strength. This is reasonably true for properties parallel to the fiber direction, but it does not hold for transverse or shear properties although the error is small for small corrections.

3   CHARACTERIZATION OF LAMINA AND LAMINATE PROPERTIES

The characterization of lamina and laminate properties for composite materials is very complex with an array of test methods available to determine a single property in many instances. In this section the test methods that can be used to characterize specific properties are defined. In cases where more than one method is discussed, the differences are clearly brought out and guidelines are given about the use of each method.

3.1   Tension Testing

Tension testing is about the simplest composite test to perform and to understand, yet there are many ways to perform tension tests on composites. To characterize the unidirectional lamina, the test can be either $0°$, $90°$, $\pm 45°$, or off-axis. The first two measure the tensile properties in the 1 and 2 material directions, respectively. The $\pm 45°$ and off-axis methods measure shear properties through tensile loading. The test can also be performed on laminates, woven fabrics, or discontinuous fiber composites to measure effective tension properties. When characterizing multidirectional laminates, we can test with or without tabs.

Discussion of the lay-ups that produce shear loading are deferred to Section 3.4 so as not to confuse the various aspects of tension testing with shear characterization. Since the tabbed versus untabbed specimen issue is important, it is covered in this section, and the results will apply to all subsequent test methods that employ tabbed specimens.

The tension test is used to measure the tensile strength, tensile modulus, Poisson's ratio, and ultimate strain to failure for composite lamina or laminates. The most common method employed for tension testing is ASTM D3039 [10]. There are several other methods; for example, the dogbone tension test, the streamline tension test, and ASTM D638. The relative characteristics of these methods are discussed in a report by Oplinger and co-workers [17]. All use basically the same procedure but employ different specimen designs. The ASTM D3039 employs a straight-sided specimen with beveled tabs. Most of the other designs use specimens that have the test section cross-sectional area reduced by routing or grinding down the width, thickness, or both. The benefit of these designs, if there is one, rarely justifies the extraordinary cost penalty for fabrication; an exception is the testing of ultrahigh modulus fiber systems. However, for most composite systems there has been no documented improvement in tension properties with the use of reduced section specimen designs. The D3039 method is discussed here since it is the most commonly used for unidirectional composites and composite laminates. The D638 [11] method is used for discontinuous fiber composites, that is, molding compounds.

3.1.1   Tension Specimen Design

Typical geometry and dimensions are given in Figure 4.12 for the straight-sided tension specimen. The $0°$ tension specimen is nominally 23 cm long, about 6 plies thick, and cut to a 1.27-cm width, while the $90°$ specimen is 2.54 cm wide and at least 8 plies thick. End tabs should be used on lamina and are nominally 3.8 cm long, 1.6–3.2 mm thick, and beveled with an angle between $15°$ and $30°$. Tabs should be made from [0/90] glass/epoxy or woven cloth laminates. Printed circuit board is often used because of its high thickness tolerances. Laminates and sheet molding compound use the same geometry as the $90°$ specimen and have the thicknesses defined by laminate configuration or fundamental sheet thickness. Laminates and sheet molding compound can be tested with or without tabs, although tabs are recommended for thin specimens. Specimens can be scaled up or down if needed, and for some materials longer test sections are employed. As a very general rule, the length of tension specimens is normally greater than or equal to eight times the specimen width.

Quality of the machined edges is important, especially for $90°$ tension

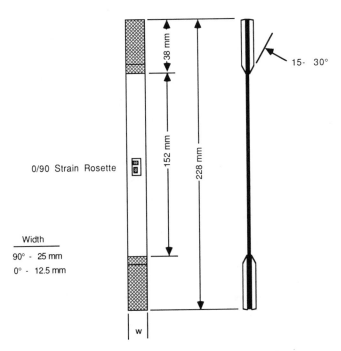

Figure 4.12   Straight-sided D3039 [10] tension specimen.

specimens. Specimens should also have dimensions within the tolerances specified for the test. Thickness uniformity in the gripping region is of critical importance.

3.1.2   End Tab Considerations

End tabs provide a method for load introduction through shear while protecting the specimen from damage. The tabs are recommended to prevent damage on lamina that have poor transverse strength but can sometimes be omitted on laminates, especially thick laminates. Properly installed, tabs improve gripping uniformity by providing a smoother, truer surface. When tabs are omitted, emery cloth is often used to improve uniformity of gripping.

To function properly, tabs should be strong enough to transfer load into the specimen and have elastic properties similar to the material under test. The additional thickness of the tab causes a stress concentration to be generated at the tab-test section transition. Figure 4.13 from Oplinger and co-workers [17] shows the stress concentration for three tab bevel angles as a function of the distance away from the tab. The bevel angle is designed to minimize the stress concentration; and, as shown in Figure 4.14, lower angles provide lower stress concentrations. The stress concentration must be minimized so that failure will

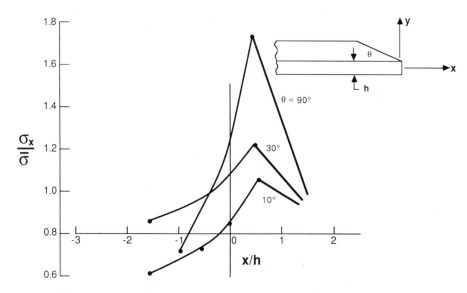

Figure 4.13   Stress concentration due to load introduction through tabs [17].

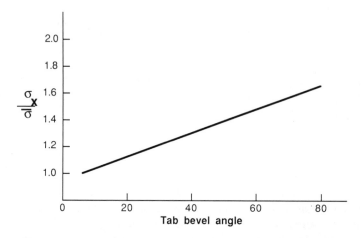

Figure 4.14   Peak stress concentration as a function of bevel angle [17].

occur in the test section where the state of stress is more uniform. Tab bevel angles cannot practically be less than 5°, since the tabs begin to take too much of the test section.

Recently, research on tension testing of composites reinforced by intermediate modulus, high strain–failure carbon fibers (fibers with 275-GPa modulus and 5500-MPa strength) has shown that the omission of tabs can result in higher measured strengths. These results are obtained when using hydraulic grips in conjunction with emery paper. The results are not thoroughly understood and the inverse trend holds true for most composite systems, especially with moduli less than 235 GPa.

3.1.3 Test Procedure

Tension testing is performed in a universal test machine using wedge action grips. Modulus can be measured either with an extensometer or a strain gauge mounted longitudinally on the specimen. If Poisson's ratio is desired, a 0°/90° strain rosette should be employed. The specimen is tested monotonically to failure while recording load, crosshead displacement, and strain. Normally, the test is run at a crosshead speed of 2 mm/min. Failure mode and location should be noted for each test along with ultimate load–failure. A failure located outside the test section justifies rejection of the result. Some typical failures for 0° and 90° carbon fiber composites are shown in Figure 4.15. The 0° carbon specimens literally explode. Safety glasses and a protective shield are recommended during tension testing.

From the ultimate load at failure the ultimate tensile strength is determined as the load divided by the cross-sectional area in the test section

$$\sigma_t^{ult} = P/A \tag{7}$$

where σ_t^{ult} is the ultimate tensile strength of the composite, P is the load at failure, and A is the cross-sectional area.

The modulus is determined as the slope of the stress–strain curve. Current trends are leaning toward defining the modulus as the slope between two strain

Figure 4.15 Typical failures of 90° (a) and 0° (b) carbon fiber composites.

(a)   (b)

levels, such as 1000 and 6000 microstrain

$$E_t = (\sigma_{6000} - \sigma_{1000})/(\varepsilon_{6000} - \varepsilon_{1000}) \tag{8}$$

where E_t is the tensile modulus and the σ and ε terms represent stress and strain at the train levels indicated by the subscripts.

Poisson's ratio is determined as the ratio defined by

$$v = -\varepsilon_2/\varepsilon_1 \tag{9}$$

where v is Poisson's ratio, ε_1 is the longitudinal strain, and ε_2 is the transverse strain.

3.1.4   Testing Issues

The tension properties of lamina are sensitive to fiber volume fraction and fiber–matrix interface. The transverse tension strength is particularly sensitive to the quality and uniformity of interfacial bonding. To be used on a comparative basis, tension test results must be normalized to some standard reference volume fraction. A linear interpolation is used to do this. The linear interpolation is quite accurate for 0° properties. It does not hold for transverse properties, but for small corrections in volume fraction the error is usually accepted.

The properties obtained for laminate tension tests are effective properties. The test method works when laminates are balanced and symmetric, but problems occur when unsymmetric and/or unbalanced laminates are tested. For these cases bending extensional coupling and shear coupling effects can produce distortions in the state of stress in the test section. Strain gauges mounted on the front and back surfaces of the test specimen can allow the measurement of modulus, but little can be done to compensate for the coupling effect on ultimate strength.

Discontinuous fiber systems can form pseudolaminated structures when material flow during molding produces a skin-core effect. With the skin-core effect, material flow during processing produces a highly oriented layer near the mold surface. This layer possesses different properties than those of the core, which has a more random fiber orientation. If skins of different thickness exist, an unsymmetric laminate is formed and the specimen must be treated as such. A more common problem resulting from the skin-core effect is the impact of machining on properties. Machined and molded tensile bars of the same material can produce significantly different results. Discontinuous fiber composites have also been shown to be strain-history dependent. When loaded past a certain elastic limit, the stress–strain behavior changes.

Woven fabric composites behave similar to laminates. The main difference is the level of homogeneity. When using strain gauges on woven fabric materials, the strain gauge size must be selected to average deformation over a representative portion of the fabric structure. Failure to use a sufficiently large strain gauge will result in large variability in the apparent strain behavior.

3.2   Compression Testing

Compression tests measure the compression response of composite materials; compression response is an effective structural property, not an intrinsic material property. As much as materials formulators would like to be able to measure an intrinsic compression property for composite materials, no test method exists that will accomplish this.

The main problem with compression testing is the stability of the test specimen. Under compression loading there is a tendency for the specimen to buckle out of plane at some critical buckling load. This response is governed by the elastic modulus of the column, the column length, the moment of inertia of the column, and the fixity condition at the ends of the column. For end loading without supports the Euler column buckling equation for a column pinned at both ends applies and the critical buckling load predicted is

$$P_{cr} = \pi^2 EI/L^2 \tag{10}$$

where P_{cr} is the critical buckling load, L is the column length, E is the axial Young's modulus, and I is the moment of inertia for the column.

When a shear loading technique is employed for load introduction, the fixity condition is better approximated by assuming that both ends of the column are clamped, which changes the critical buckling load to

$$P_{cr} = 4\pi^2 EI/L^2 \tag{11}$$

and gives a fourfold increase in load carrying capacity before buckling. In reality the fixity is not perfectly clamped but lies somewhere between pinned and clamped—closer to the clamped condition. For design, the conservative approach is to use (10). By making the test section length shorter than the length that will produce critical buckling prior to the ultimate failure load of the composite, we can measure an ultimate strength.

There are many test methods for compression testing, each with its particular specimen design. The methods fall into one of three categories according to how the load is introduced into the test specimen; that is, end loading, shear loading, and flexural sandwich loading. In the end-loaded category are ASTM D695 [18] and several modified derivatives, the Royal Aircraft Establishment (RAE) method, the Lockheed method, the Northrop method, the Southwest Research Institute method, and the end-loaded sandwich beam method. The shear-loaded methods include the Celanese and the Illinois Institute of Technology Research Institute (IITRI) methods described as part of ASTM D3410 [19]. The flexural method involves four-point bending of a thin-skinned sandwich beam [20], also described in method D3410 [19]. One method for each of these loading schemes is discussed since it covers most of the salient features of all the methods in its class.

Regardless of the method, ultimate compression strength, compression modulus, and ultimate compressive strain–failure are the primary properties measured by the compression test. These properties can be determined on unidirectional lamina, cross- and angle-ply laminates, discontinuous fiber composites, and woven fabrics.

3.2.1. End-Loaded Compression Testing

End-loaded tests are very popular because they require the simplest fixturing of all the compression tests. The RAE test (Figure 4.16) simply bonds or clamps a small section of laminate into aluminum blocks that are then placed between two parallel loading plates in a universal test machine and loaded to failure. The ASTM D695 method [18] as modified by Boeing is probably the most commonly used end-loaded compression test. The fixture shown in Figure 4.17 supports the specimen shown in Figure 4.18 while it is end-loaded. For strength measurement tabs are bonded onto the specimen, increasing the bearing area in the end-loaded regions. For modulus determination the specimen is not tested to failure and the tabs are omitted. Strain gauges are mounted on both surfaces, and the strain is averaged to obtain the stress–strain curve. The strength specimen is supported by the fixture along its whole length except in the short

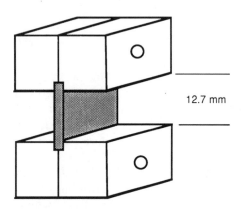

12.7 mm

Figure 4.16 Royal Aircraft Establishment (RAE) test specimen and fixture.

Figure 4.17 Modifed ASTM D695 [18] compression test fixture.

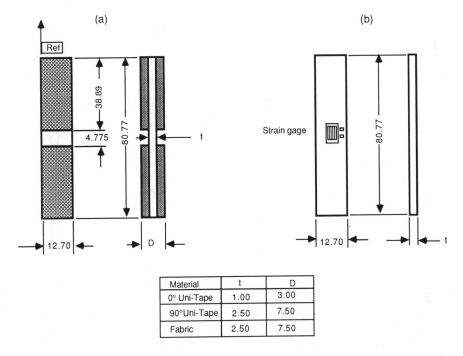

Material	t	D
0° Uni-Tape	1.00	3.00
90°Uni-Tape	2.50	7.50
Fabric	2.50	7.50

Figure 4.18   Modified ASTM D695 [18] test specimen details (*a*) strength measurement and (*b*) modulus measurement. (*Note*: All tolerances are ±0.1 mm. Tabs must be flat, parallel, and of equal thickness to within 0.05 mm. Ends must be surface ground flat and perpendicular to the reference axis to within 0.025 mm.)

4.8-mm test section. Other methods, like the Southwest Research Institute method, support the entire gauge section of the test specimen (Figure 4.19).

While the fixturing is simple for end-loaded tests, load introduction can cause axial splitting or bearing failures on the loaded ends. This must be prevented and has prompted the use of tabs, potting compound application, or fixture supports at the specimen ends. Another problem is alignment. Alignment depends on the machining precision of the specimen ends; that is, on the parallelism between opposing ends and the perpendicularity of these ends to the specimen loading axis. Not only must the specimen be precise, but the test machine platens must also be precisely parallel and aligned with the axis of the test machine. The third source of alignment error is specimen insertion. Much of the fixture design complexity revolves around the preservation of specimen alignment in the test machine.

3.2.2   Compression Tests Using Shear Load Introduction

There are basically two test methods that use shear load introduction to achieve compression loading: the IITRI and Celanese methods. The IITRI method

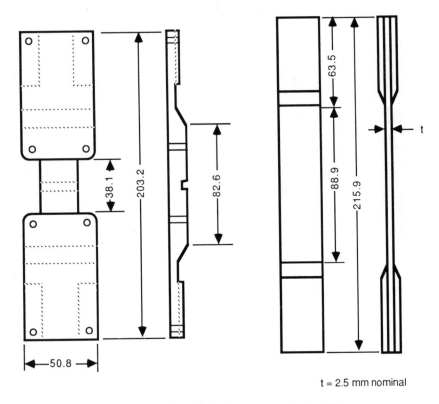

t = 2.5 mm nominal

Figure 4.19 Southwest Research Institute compression test fixture.

shown in Figure 4.20 employs trapazoidal-shaped wedge action grips that seat on compression in massive upper and lower fixturing blocks that are aligned by two steel dowels. The typical unidirectional specimen geometries employed with this test are shown in Figure 4.21. Thickness is not tightly restricted, and specimens of other widths can be tested, up to the limits imposed by the geometry of the grips. The original fixture design incorporated wedges capable of 2.5-cm-wide specimens but modified fixtures were fabricated that are capable of handling 5.0-cm or even wider specimens. This feature is very useful for testing laminates.

The Celanese method employs conical-shaped wedges that are seated in cylindrical blocks constrained and aligned by a steel pipe as shown in Figure 4.22. The test is designed primarily to measure lamina properties with the specimen geometry shown in Figure 4.23. The thickness dimension specified in this test is critical, which restricts the use of this method for laminates.

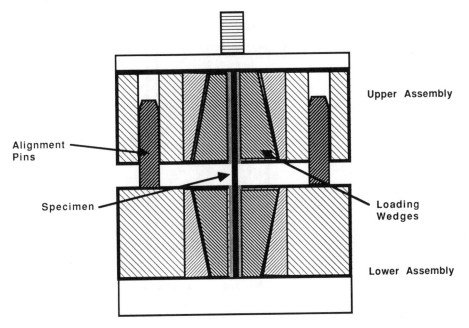

Figure 4.20 Illinois Institute of Technology Research Institute compression test fixture.

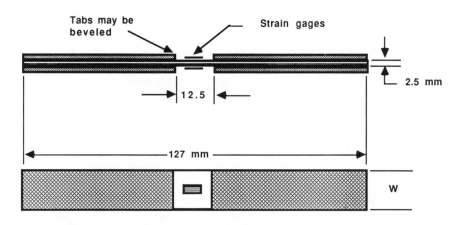

W = 12.5 mm. (90° Specimen)

W = 6.4 mm. (0° Specimen)

Figure 4.21 Typical IITRI test specimen design.

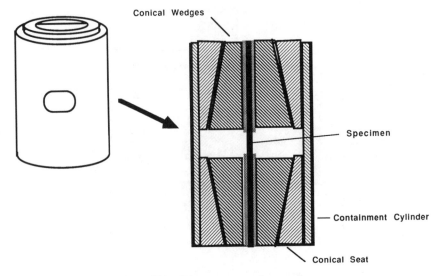

Figure 4.22   Celanese compression test fixture.

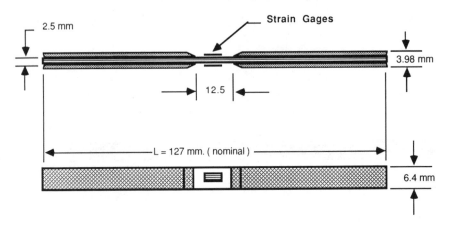

Figure 4.23   Specimen dimensions and geometry for Celanese method.

Specimens must be precisely ground to this thickness so that the grips will seat properly. Failure of the grips to seat properly results in reduced apparent strength and larger scatter in the data. Recently, constant radius cylindrical wedges have been developed using the Celanese fixture shell, eliminating the need for precise specimen thickness control. The test then functions like the IITRI but with less massive fixturing for alignment.

The test methods produce comparable results with properly prepared specimens and careful fixture alignment. The IITRI fixturing allows a much broader range of specimen geometries to be tested, but it is very bulky. The flexibility in specimen geometry makes this method useful for laminate testing. The bulk of the fixture makes it very difficult to use at elevated temperatures. Binding of the alignment posts can even occur at temperatures exceeding 250°C. The Celanese fixture allows only a single specimen geometry, but it is smaller and easier to use at elevated temperature. It, too, is subject to binding at temperatures above 250°C.

3.2.3 Sandwich Beam Method

The sandwich beam method is based on the four-point bending of a sandwich beam composed of two thin composite face sheets bonded to a high density honeycomb core. The core acts as a shear web allowing the tension and compression loads to be reacted almost exclusively by the face sheets. Since the face sheets are thin, the deformation is approximately uniform through the thickness.

Typical construction of the sandwich beam compression specimen is shown in Figure 4.24. The tensile face can be a high strength material like titanium or a thicker section of the same material used on the compression face. Assuming both faces are made out of the same material, the compression strength is determined by

$$\bar{\sigma} = \frac{PL}{4bh(2t + H + h')} \tag{12}$$

where $\bar{\sigma}$ is the average stress in the face sheets neglecting any contribution of the core, P is the load, L is the span between the outer load reaction points, b is the beam width, h is the skin thickness of the compression face, h' is the skin thickness of the tensile face, and t is the core thickness. Modulus is determined using similar assumptions from the relationship

$$E_x = \frac{PL}{4bh\varepsilon_x(2t + h + h')} \tag{13}$$

where E_x is the longitudinal compression modulus, ε_x is the longitudinal strain, and the other terms are as previously defined.

To prevent failure of the face sheets under the load noses, rubber pads are normally employed. Otherwise the test is conducted just like a four-point bending test. The primary drawbacks of this test are the large specimen size and expensive fabrication procedure that are required.

3.2.4 Test Procedures

Once inserted in the fixturing, the test specimen is loaded monotonically to failure. Ultimate load is recorded and used to determine the apparent compressive strength of the materials. If modulus is required, the strain must be recorded

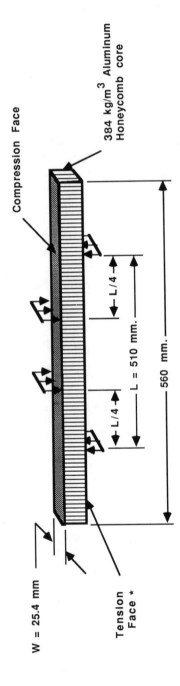

Figure 4.24 Sandwich beam compression test specimen.

as a function of load during the test. In the IITRI and Celanese tests strength and modulus are determined simultaneously during the test. The modified D695 method [18] employs separate specimens for strength and modulus measurement. The modulus specimen is loaded to only 25% of its ultimate strength.

A compressometer can be used to measure strain, but back-to-back strain gauges are preferred. Strain gauges mounted on each surface provide a direct measure of specimen bending or instability and should be plotted independently. Modulus is determined by averaging the two strains over the linear portion of the stress–strain curve.

The strains from the opposing specimen surfaces should be used along with failure mode examination to interpret test results. For a well-aligned, precisely machined specimen the strains from both surfaces will be in close agreement, producing curves similar to those labeled A in Figure 4.25. Bending will cause the strains to diverge with increasing load until instability causes strain reversal, illustrated by the curves labeled B in Figure 4.25. Interestingly, the bending behavior does not significantly affect modulus as determined by averaging the two strains, but apparent strength is decreased.

The failure mode produced by a compression test should always be noted. Typical compression failures for unidirectional laminates are illustrated in Figure 4.26. Often failures will propagate into the tab section, but as long as the failure is not solely in the tabs, the result is considered acceptable. As seen in

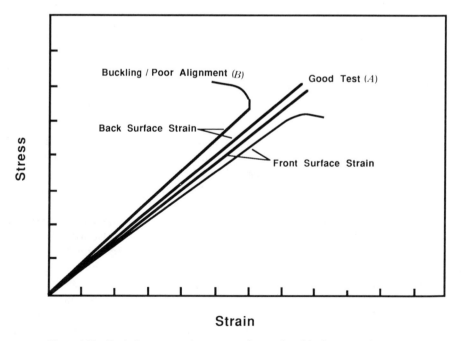

Figure 4.25 Typical stress–strain responses for good and bad compression tests.

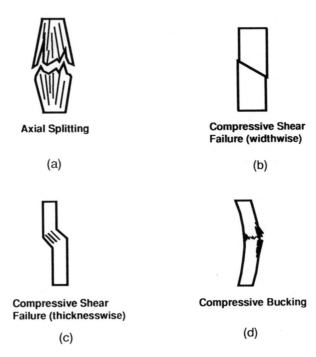

Axial Splitting

(a)

Compressive Shear Failure (widthwise)

(b)

Compressive Shear Failure (thicknesswise)

(c)

Compressive Bucking

(d)

Figure 4.26 Typical failure modes for compression specimens.

Figure 4.26a–c, compression produces the shear type of failure common in metals along with some axial splitting. Buckling-induced failures exhibit the compression kinking on one side with the flexural-type tension failure on the opposing surface as shown in Figure 4.26d.

3.2.5   Summary

Many compression test methods are available, but they can be broken down into three basic types: end-loaded, shear-loaded, and the sandwich beam bending tests. The sandwich beam bending test is material intensive and expensive and is not used routinely for compression characterization. The D695 type test is very popular for materials screening since it produces consistent strength and modulus measurements. The main drawback is the requirement for separate strength and modulus specimens. The IITRI test is also widely used for lamina and laminate characterization for both materials screening and design data generation. The Celanese test is effective but less widely used because of the specimen limitations. While the tests have been described in the context of continuous fiber laminates, the end-loaded and shear-loaded test methods are also suitable for testing discontinuous fiber composites. As with testing laminates, 2.5-cm widths should be used as a minimum when characterizing discontinuous fiber composites.

3.3 Flexure Testing

Flexure tests measure the bending properties of composite materials. The bending response of a composite system is a structural response; hence, the flexural properties are not the same as those measured by a pure tension or pure compression test. Because the specimen is small and easy to fabricate the flexural test is extremely useful for comparative studies of material performance and for quality control. At elevated temperatures and under hot–wet conditions, the flexural test is often the only method available to measure properties.

Flexure is used to measure apparent strength and stiffness of composites. For unidirectional laminates, flexural strength and stiffness can be directly related to the tension and compression properties. Lamina properties can be characterized in both the longitudinal and transverse directions. For laminates, effective bending strength and stiffness is measured and is a function of ply orientation and stacking sequence.

There are several methods for performing flexural testing that are distinguished by the number of load reaction points. Three-, four-, and five-point bending methods have been reported in the literature [21], but the most common are the three- and four-point bending methods and only these two methods are covered.

3.3.1 Background

Flexure, or bending, of a beam produces a complex state of stress in the beam. Ignoring any stress concentrations at the load reaction points and assuming the beam is homogeneous, the bending moment creates bending stresses that are maximum at the outer surface fibers and zero at the neutral surface. The stresses are tensile on one surface and compressive on the opposing surface as shown in Figure 4.27. Simultaneously, a parabolic distribution of shear stress is generated across the thickness of the beam, in those regions where a shear force is acting. Depending on the length-to-thickness aspect ratio chosen for the test and the

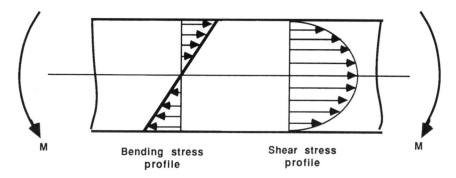

M

**Bending stress
profile**

**Shear stress
profile**

M

Figure 4.27 Bending and shear stress distributions in beam.

relative magnitudes of the tensile, compression, and shear ultimate strengths, the specimen may fail in tension, compression, or shear.

The shear and bending moment diagrams used to determine stresses in the beam for three- and four-point bending are shown in Figures 4.28 and 4.29. The fundamental difference in these two methods is clear from these diagrams. For the four-point bending test, the shear force is zero between the two inner load reaction points and the moment remains constant, while the moment for the three-point bending test is changing throughout the test section. The data reduction formula for strength and modulus reflects these distinctions. Also, the practical ramifications in terms of displacement measurement or strain gauge location will be apparent in Section 3.2.2.

The homogeneous material assumption is valid for lamina, even though they are orthotropic. Laminates, however, are not homogeneous, and the bending stresses in a laminate depend on the ply orientations and stacking sequence as illustrated in Figure 4.30. The stress concentrations at the load reaction points can be significant enough to influence the test results. This is especially true for composites with poor transverse compression strength. The test should be designed to minimize the effects by adjusting the span-to-depth ratio and/or the type of binding setup used. Increasing the span lowers the load required to induce failure. Comparing the load required to generate a constant failure stress for the three-point bending and four-point bending with inner spans of $L/3$, and the four-point bending with an inner span of $L/2$, Figure 4.31, shows that four-point bending with an $L/3$ span setting requires the lowest load at any given span-to-depth ratio.

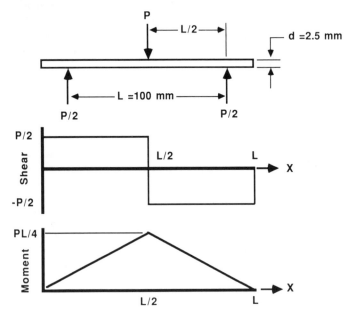

Figure 4.28   Shear force and bending moment diagrams for three-point bending.

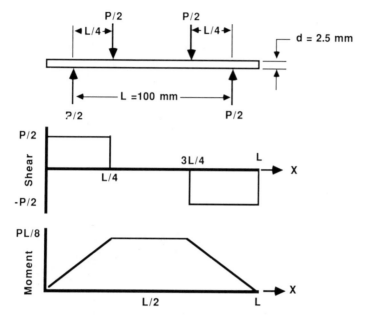

Figure 4.29   Shear force and bending moment diagram for four-point bending.

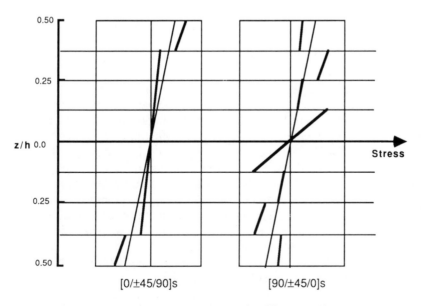

Figure 4.30   Stress distributions in laminates for different stacking sequences.

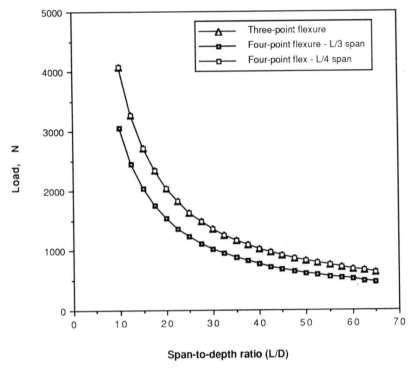

Figure 4.31   Effect of span and flexure configuration on load at failure.

3.3.2   Test Procedures

THREE-POINT BENDING

A typical three-point bending test setup is shown in Figure 4.32. The test specimen is nominally $1.27 \times 12.7 \times 0.2$ cm. The span between the outer load noses is determined according to the material being tested. Table 4.2 gives span settings for some common composite systems. Most flexure fixtures have guide pins that assist in the alignment of the specimen in the fixture. The test is conducted by loading the specimen monotonically to failure with the central load nose. Modulus can be determined by either recording the deflection at midspan or by a strain gauge mounted on the tensile face at midspan. Because of the bending moment variation along the beam, the precise location of the deflectometer or strain gauge is critical. Strain or midspan deflection is recorded as a function of load during the test.

The load versus deflection curve is recorded along with the ultimate load carrying capacity of the specimen. Assuming the material is homogeneous, orthotropic, and possessing equal apparent compressive and tensile moduli, simple beam theory is invoked to reduce the data. The strength for three-point

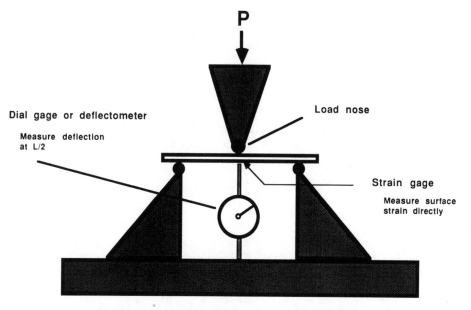

Figure 4.32   Typical three-point bending test setup.

Table 4.2   Typical Span Settings for Flexure Testing of Composite Systems

Material Strength	Span Settings	
	Modulus	Strength
Boron/Epoxy	32:1	32:1
Carbon/Epoxy	32:1	32:1
Carbon/Epoxy (90°)	16:1	16:1
Glass/Epoxy	16:1	16:1
Kevlar/Epoxy	16:1	48–60:1
Carbon/TP[a]	32:1	48–60:1
Carbon (IM)[b]/Epoxy	32:1	48–60:1

[a]TP stands for thermoplastic matrix materials.
[b]IM is intermediate modulus fiber (300–350 GPa).

bending is determined using

$$\sigma_{\text{flex}} = \frac{3PL}{2bd^2} \tag{14}$$

where σ_{flex} is the flexure strength, P is the ultimate load, L is the span, b is the specimen width, and d is the specimen thickness. The modulus (E_{flex}), neglecting shear deformation, is found from the following equation, given that the mid-span deflection is measured.

$$E_{flex} = \frac{PL^3}{4bd^3\delta} \tag{15}$$

Depending on the shear modulus and span/depth ratio, the contribution of shear deformation to the observed deflection can be significant. To correct for the contribution of shear deformation the equation for modulus becomes [22]

$$E_{flex} = \frac{PL^3}{4bd^3\delta}\left(1 + \frac{3d^2E_x}{2L^2G_{xz}}\right) \tag{16}$$

where E_x is the extensional modulus along the specimen length, G_{xz} is the interlaminar shear modulus, and the other terms were defined earlier. If strain is measured directly, modulus is determined by the usual method as the slope of the stress–strain curve. Equation (14) is used to determine the stress for any given load and the equation for finding the flexural modulus is

$$E_{flex} = \frac{3PL}{2bd^2\varepsilon_x} \tag{17}$$

where ε_x is the longitudinal strain measured on the tensile face of the specimen. The flexural stress–strain response is nonlinear and care should be taken to determine modulus on the linear portion of the curve, or at a clearly defined point.

Since the flexural response of the beam is based on the assumption of small deflection theory; that is, the deflection is less than the thickness of the beam, the flexure test is discontinued if 5% strain is reached. The span-to-depth should be reduced until the 5% limit is not exceeded. For specimens that exhibit large deflections, deflections greater than 10% of the support span, a correction for the geometric nonlinearity should be applied. This corrects for the horizontal thrust effects induced by the large deflection of the system [23, 24]. The correction for stress is

$$\sigma_{flex} = \frac{PL}{bd^2}\left[1 + 6(\delta/L)^2 - 4(d/L)(\delta/L)\right] \tag{18}$$

where the only new term δ is the midspan deflection of the flex specimen at load P.

FOUR-POINT BENDING

A typical four-point bending test setup is shown in Figure 4.33. The specimen geometry for four-point bending is the same as that used in three-point bending.

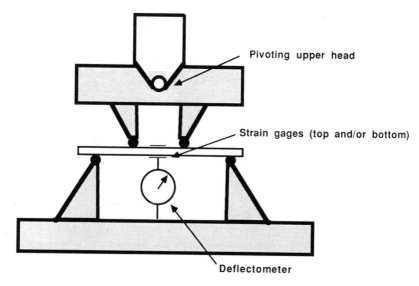

Figure 4.33 Typical four-point bending test setup.

Because there are four points of contact, a pivoting design of the upper loading assembly is recommended to assure symmetric loading by the two upper load noses. The positioning of the two upper load noses is normally set so that the span between them is one-third or one-fourth of the total span. Some tests, like the 90° flexure, are always run in four-point flex with an inner span of $L/2$. For 0° lamina and laminates, the balance of properties of the material should be considered as well as the purpose of the test. Criteria for selecting the appropriate test setup are discussed in Section 3.3.3.

The four-point test is run the same way as the three-point test, recording the load deflection and/or load–strain behavior and ultimate load at failure. Modulus can be measured using either a deflectometer or strain gauge mounted at the $L/2$ midspan location. Strain gauges can be mounted on the top and bottom surfaces to measure the apparent tensile and compression modulus that for some materials, like carbon fiber composites, are not the same. Precise location of the strain gauge is not critical for four-point bending since the moment, hence, stress in the outer fiber, is constant along the inner span. The deflectometer does need to be precisely positioned, since deflection is a function of the coordinate along the length of the beam. Crosshead motion can be used to measure deflection, but the proper equation must be used to determine modulus based on a deflection measured at the quarter span location.

Assuming the material is homogeneous, orthotropic, and possessing equal apparent compressive and tensile moduli, simple beam theory is invoked to reduce the data. The strength for four-point bending with inner spans of $L/3$ and

$L/4$ are determined using (19a) and (19b), respectively,

$$\sigma_{\text{flex}} = \frac{PL}{bd^2} \tag{19a}$$

$$\sigma_{\text{flex}} = \frac{3PL}{4bd^2} \tag{19b}$$

The modulus, neglecting shear deformation, for inner spans of $L/3$ and $L/4$ is found from the following equations given that the midspan deflection is measured.

$$E_{\text{flex}} = \frac{0.21PL^3}{bd^3\delta} \tag{20a}$$

$$E_{\text{flex}} = \frac{0.17PL^3}{bd^3\delta} \tag{20b}$$

Depending on the shear modulus and span-to-depth ratio, the contribution of shear deformation to the observed deflection can be significant. To correct for the contribution of shear deformation the equation for modulus for loading at the quarter span becomes

$$E_{\text{flex}} = \frac{0.17PL^3}{bd^3\delta}(1 + 0.125S) \tag{21}$$

where

$$S = \frac{3d^2 E_x}{2L^2 G_{xz}}$$

If strain is measured directly, modulus is determined from the slope of the stress–strain curve. Equation (19) is used to determine the stress for any given load and is divided by the measured strain. The flexural stress–strain response is nonlinear and care should be taken to determine modulus on the linear portion of the curve. If another approach is taken, it should be clearly defined.

Since the flexural response of the beam is based on the assumption of small deflection theory, the test is discontinued if 5% strain is reached. The span-to-depth should be reduced until the 5% limit is not exceeded. For specimens that exhibit large deflections (deflections greater than 10% of the total span) the correction for the elastic nonlinearity should be applied. This corrects for the horizontal thrust effects induced by the large deflection of the system. The corrections for stress are

$$\sigma_{\text{flex}} = \frac{PL}{bd^2}[1 + 4.7(\delta/L)^2 - 7.04(d\delta/L^2)] \tag{22a}$$

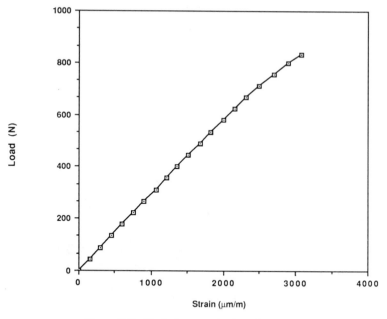

Figure 4.34   Typical test results for a flexure test.

$$\sigma_{\text{flex}} = \frac{3PL}{4bd^2}\,(1 - 10.91d\delta/L^2) \tag{22b}$$

Typical results for a flexural test are given in Figure 4.34.

3.3.3   Testing Configuration Selection

The flexure test is a simple test using a small, low-cost specimen that can provide useful information about the relative performance of materials. It should not be used as a replacement for compression and tension testing to determine lamina properties. The selection of a particular form of flexure test is governed by the properties of the material being tested and the purpose for conducting the test.

As shown in Figure 4.31, each test loads specimens to different levels to obtain failure at a given span-to-depth ratio. The span-to-depth ratio can be adjusted to vary loading and more importantly, the ratio of shear-to-flexural stress can be developed in the beam. Figure 4.35 shows how the shear stress for each of the three common test configurations varies as a function of L/d. Figure 4.36 shows the critical span at which shear failure will occur as a function of the ratio of tensile-to-shear strength. Each configuration has a different response curve. Thus, the test configuration and span-to-depth ratio must be designed for the material being tested. The test can be designed to generate tensile or shear failures depending on the design selected.

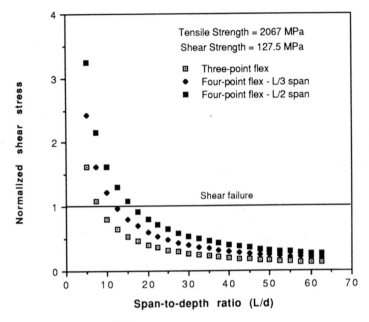

Figure 4.35 Shear stress as a function of L/d.

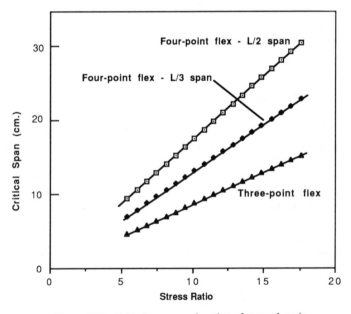

Figure 4.36 Critical span as a function of strength ratio.

A useful way to use the flexure test for materials screening and quality control is to design the test near the transition zone on one of the curves. The test will then fail in shear if the material is substandard and in tension if it is good. The flexure test is also very sensitive to compression properties, and it will fail in compression if the compressive properties are poor. One cause of poor compression properties is a poor fiber matrix interface, and so the results of the flex test can detect unusual material behavior resulting from poor fiber wetout or matrix–fiber bonding.

When reporting flexure test results it is important to report the failure characteristics or mode. The test is usually designed to produce a tension failure in the outer fibers. As discussed previously other types of failure can occur and can provide useful information about a material system.

Flexure of laminates produces only an apparent flexural strength and stiffness. These properties are dependent on ply orientation and stacking sequence. Only balanced and symmetric laminates should be tested since bending and twisting coupling causes the laminate to deform out of the plane of the test. This will produce nonuniformities in the specimen state of stress and deformation that cannot be handled using the procedures discussed here. Discontinuous fiber composites can be considered orthotropic sheets, like lamina, if the fiber orientation state is uniform through the thickness. If a layered microstructure exists, it must be treated like a laminate.

3.3.4 Testing Issues

The data reduction commonly employed for flexural testing assumes that the apparent tension and compression moduli of the material under flexure are equal. This is not always true; carbon/epoxy is a common exception. When this is the case the bending stress profile changes as shown schematically in Figure 4.37. From equilibrium we know that this results in a shift of the neutral axis and the apparent flexural modulus (E_{flex}) will be related to the tension modulus through the following relation.

$$E_{flex} = \frac{2E_t}{1 + \sqrt{E_t/E_c}} \tag{23}$$

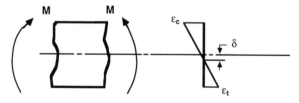

Figure 4.37 Change in bending stress distribution because of unequal moduli in tension and compression (δ = distance that the neutral axis is shifted).

where E_t and E_c are the tensile and compressive Young's moduli, respectively. This means that if E_t is less than E_c, then the apparent flexural modulus will be greater than the tension modulus; if E_t is greater than E_c, then the apparent flexural modulus will be less than the tension modulus. If E_t is equal to E_c, then the flexural and tension moduli are equal.

Strength of brittle composite materials is governed by Weibull statistics, which are a weak link type of model. Therefore, strength is a volumetric quantity and is subject to a size effect. In tension of a lamina, the volume of material in the test section is uniformly stressed. In flexure the outer fiber layers are subjected to the highest stress and therefore only a small volume of the total specimen volume is highly stressed. For this reason strength in flexure may be 20–50% higher than the tension strength of the same material. Tougher, less brittle systems, such as thermoplastics, do not exhibit this behavior. Often the failure is governed by the compression strength in both the transverse and longitudinal directions, and the failure occurs at a lower value than either the tension or compression strength.

3.3.5 Summary

Flexure is a useful test for materials screening and quality control. Results from flexure may not quantitatively translate to the equivalent lamina properties measured by tension and compression tests separately. There are several configurations of flexural testing and the configuration in combination with the span should be determined to best suit the purpose of conducting the test based on the properties of the material being evaluated.

3.4 Shear Test Methods

In an orthotropic composite material shear can be measured in three material planes, the 1-2, 1-3, and 2-3 planes, as shown in Figure 4.38. There are many shear test methods; some can only measure shear in one plane, and others can measure shear in all three. Some of the methods only work for continuous fiber composites.

As with any characterization method, the utility of shear characterization methods depends not only on the quality of the results, but also on practical issues such as specimen fabrication, ease of performance, and cost. More than with any other test, these factors influence the choice of a shear test method.

The torsion test is recognized as the method that produces the purest state of shear stress. It is perfect for circular shaped pultruded materials. For laminates, on the other hand, fabrication of a tube is difficult and the change in processing conditions is likely to change properties. Other methods are better suited to laminates. For continuous fiber composites, the in-plane shear properties can be determined by either a $\pm 45°$ tension test (ASTM D3518 [25]), a two or three rail shear test as described in ASTM D4255 [26], a 10° off-axis tension test, or a V-notched beam shear test (Iosipescu or asymmetric four-point bending). Discontinuous fiber composites have fewer options: the rail shear method and

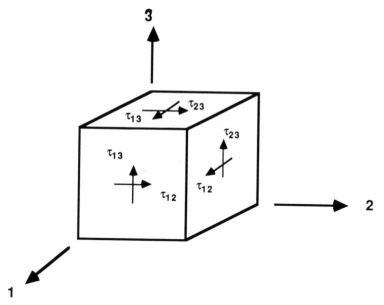

Figure 4.38   Definition of shear planes.

the V-notch beam shear method. Interlaminar properties, those in the 1-3 and 2-3 planes, can be measured by the short beam shear test (strength only), the V-notched beam shear test, a four-point bending shear test, and a notched tension test, as described in ASTM D3846 [27].

Some brief comparative comments are given on the use of the preceding methods in Table 4.3. Lee and Munro [28] devised a weighting method to evaluate several of the most popular shear test methods from an accuracy as well as from a practical viewpoint. Their results showed that the V-notched beam shear was the best all-around method, followed by the $\pm 45°$ tension test for inplane properties only.

3.4.1   Test Method Descriptions

In this section we discuss fixturing, specimen fabrication, and characteristics of several of the most used shear test methods. This should provide guidance in the selection of the correct test method for a given materials application.

TORSION TEST

The torsion test exists in two forms: torsion of a solid rod and torsion of a thin-walled tube. The test is good for measuring both shear strength and shear modulus. Solid rod torsion is useful for pultruded rods, but it is not recommended for normal sheet-based composite systems since other test methods are better suited for sheet materials. The solid rod specimen is tested in torsion by holding one end stationary and by rotating the opposite end relative to the

Table 4.3 Summary of Shear Test Methods

Test Method	Materials	Properties	Advantages	Disadvantages
Torsion	Continuous fiber tubes/rods	Strength and modulus, 1-2 plane	Accurate/true shear stress	Specimen hard to fabricate with properties equivalent to laminate
±45° Tension	Continuous fiber	Strength and modulus, 1-2 plane	Simple, accurate lamina modulus	Strength slightly inaccurate
10° Off-axis tension	Continuous fiber	Strength and modulus, 1-2 plane	Accurate modulus; good for metal matrix composites	Strength measurement poor
Rail shear	Continuous fiber lamina/laminates	Strength and modulus, 1-2 plane	Measure effective laminate shear properties	Large specimen, hard to get good strength results
	Discontinuous fiber composites	Strength and modulus, 1-2 plane	Measure inplane shear properties in discontinuous material	
Short beam shear	Continuous fiber laminates	Strength, 1-3 plane	Simple, quick quality control test for interface quality	Failure not pure shear, no modulus
Four-point flexure	Continuous fiber laminates	Strength, 1-3 plane	Simple, quick quality control test for interface quality	Larger specimen than short beam shear; similar results
V-notch shear	Continuous fiber lamina/laminates	Strength and modulus, all planes	Small specimen, simple, measure shear properties for many material types, directions	Sensitive to machining of notch, location of strain gauges; will not handle coarse fabrics
	Discontinuous fiber composites	Strength and modulus, all planes	Small specimen, simple, measure shear properties for many material types, directions	Sensitive to machining of notch, location of strain gauges; will not handle coarse fabrics

stationary end while measuring the angle of twist and torque. For better accuracy a strain gauge rosette can be used to measure shear strain in the outer fibers as a function of torque. One drawback of this test is that the shear stress is a function of radial position in the material.

A thin-walled cylinder helps to minimize the strain gradient through the wall thickness, but it is difficult to fabricate high-quality tubular specimens from prepreg. This test method is well suited to filament-wound tubular products. To work properly the material must be in the form of an orthotropic laminate (A_{16} and $A_{26} = 0$) with symmetric lay-up. The shear stress is then determined from

$$\tau_{xy} = \frac{T}{2\pi R^2 h} \tag{24}$$

where T is the applied torque, h is the cylinder wall thickness, and R is the mean radius of the cylinder. The shear strain can be found from the angle of twist per unit length by

$$\gamma_{xy} = R\phi \tag{25}$$

where γ_{xy} is the shear strain, ϕ is the angle of twist per unit length, and R is the cylinder radius. A more accurate measure of shear strain is possible from a rectangular, three-element strain rosette. The strain gradient in the thin-wall tube can be approximated by a linear relation of the shear strain γ_{xy}^0 measured on the outer surface

$$\gamma_{xy} = \left(\frac{1 + r}{R}\right) \gamma_{xy}^0 \tag{26}$$

where γ_{xy} is the shear strain, γ_{xy}^0 is the uncorrected shear strain, and r is the radial position through the thickness of the cylinder wall. The use of gauges on the inner and outer surfaces of the tube can verify the accuracy of this relationship.

Tubular specimens should have a ratio of radius-to-wall thickness, R/h, of greater than or equal to 10 to minimize the strain gradient through the wall. The gauge length should be such that L/R is greater than or equal to 8, where L is the tube length between the grips. The wall thickness and tube radius must be sized such that torsional buckling cannot occur.

±45° TENSION TEST

The simplest and most reliable shear test is the ±45° tension test. It can be used to measure the shear strength, shear modulus, and ultimate shear strain to failure of continuous fiber materials. It is not applicable to discontinuous fiber materials since the ±45° orientation cannot be developed in those materials.

The test uses an 8-ply ±45° symmetric laminated test specimen fabricated to the geometry shown in Figure 4.39. The specimen is instrumented with a two-element strain rosette as shown and tested in tension to ultimate failure while

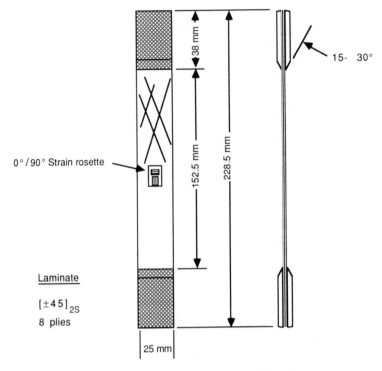

0°/90° Strain rosette

Laminate

$[\pm 45]_{2S}$

8 plies

38 mm

152.5 mm

228.5 mm

15- 30°

25 mm

Figure 4.39 Specimen geometry for the $\pm 45°$ tension test.

recording the load versus strain response. Because of the $\pm 45°$ lay-up, the shear stress, τ_{12}, is one-half the tensile stress.

$$\tau_{12} = \frac{\sigma_x}{2} \qquad (27)$$

where

$$\sigma_x = \frac{P}{wt}$$

and P is the ultimate load, w is the specimen width, and t is the thickness. While a complex state of strain is developed in the specimen, the shear strain is found to be simply

$$\gamma_{12} = \varepsilon_x^0 + \varepsilon_y^0 \qquad (28)$$

where γ_{12} is the shear strain, and ε_x^0 and ε_y^0 are the measured strains in the x and y directions on the specimen. Figure 4.40 shows a typical shear stress–shear strain response for a carbon fiber composite material. The modulus determined

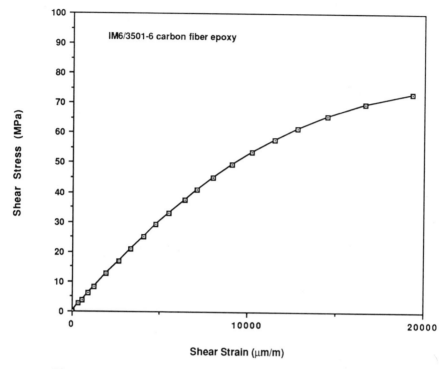

Figure 4.40 Typical shear stress–shear strain curve for the $\pm 45°$ tension test.

by this test is very accurate. The strength is influenced by the angle ply lay-up. Tension and compression tests give different strengths because the transverse stress component influences the strength [29].

10° OFF-AXIS TENSION TEST

The 10° off-axis test uses the principle that an off-axis ply produces shear coupling. The shear stress (τ_{xy}) is

$$\tau_{12} = mn\sigma_x \tag{29}$$

where $m = \cos \theta$ and $n = \sin \theta$, and the shear strain is found from

$$\gamma_{12} = -(m^2 + 2mn - n^2)\varepsilon_x - (m^2 - 2mn - n^2)\varepsilon_y + 2(m^2 - n^2)\varepsilon_{45} \tag{30}$$

where ε_x, ε_y, and ε_{45} are the strains measured in the directions indicated by the subscripts and θ is the off-axis angle. The use of this equation for the off-axis test requires that we measure strains in three directions with a rectangular three-element strain rosette with gauges oriented in the x (0°), y (90°), and 45° directions. The initial tangent modulus determined for the 10° off-axis test will be similar to that measured using the $\pm 45°$ tension test, but the overall stress–

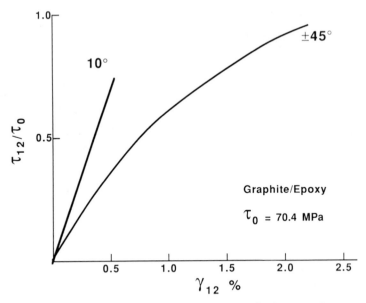

Figure 4.41 Comparison of ±45° and 10° off-axis test results.

strain response and the ultimate strength determined from the two methods are different [30]. In fact, various off-axis specimens were tried and all produce similar results; good initial tangent modulus, but ultimate strengths dependent on off-axis angle as shown in Figure 4.41.

Because of the shear coupling produced in an off-axis specimen, proper design is important to minimize the influence of shear coupling on the uniformity of stress in the test section. Under tension the specimen tends to deform in shear, but it is constrained by the grips producing a deformation shown in exaggerated form in Figure 4.42. From a mechanics analysis it can be shown that the length-to-width aspect ratio must be large to minimize error in modulus measurement. Using the approach described by Pagano and Halpin [31] we can minimize the error η in the apparent modulus by the following equation:

$$\eta = \frac{3\eta_{xy}}{\left(\dfrac{3E_x}{G_{xy}} + \dfrac{2L^2}{W^2}\right)} \tag{31}$$

where η_{xy} is the unconstrained shear coupling ratio of the extensional compliance to the shear compliance (S_{11}/S_{16}), L is the specimen length, and W is the specimen width. Since it is not practical to control the elastic properties of the material, the length-to-width aspect ratio must be increased to decrease the error. Often a 30-cm-long by 1.27-cm-wide specimen is used for this test.

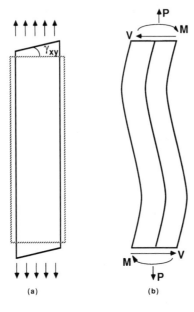

Figure 4.42 Shear coupling deformation for off-axis test specimen. The unrestrained deformation shown in (a) gives way to the deformation shown exaggerated in (b) when end constraints are applied. Reprinted from *J. Composite Materials*, 1968, Figs. 1 and 2, p. 19 © Technomic Publishing Co., Inc., 1968.

RAIL SHEAR TEST

There are two configurations for the rail shear test; they are the two-rail and the three-rail tests. The two-rail specimen is shown in Figure 4.43, and the three-rail specimen is simply the two-rail specimen extended by being reflected about the indicated plane. It should be noted that the holes in the specimen are oversized for the fasteners used to assemble the fixture so that no load will be transferred through bolt bearing. The two-rail test is conducted by mounting the specimen in the fixture shown in Figure 4.44 and loading the assembly in tension at the indicated angle until failure. To measure modulus, strain gauges must be mounted in the center of the gauge section at $\pm 45°$ to the specimen's long axis. The shear stress is calculated as

$$\tau_{xy} = \frac{P}{bh} \tag{32}$$

where b and h, respectively, are specimen length and thickness. By measuring the ultimate failure load we can determine the ultimate strength.

The shear strain is determined from the equation

$$\gamma_{xy} = \varepsilon_{+45} - \varepsilon_{-45} = 2\varepsilon_{45} \tag{33}$$

The shear modulus is normally determined from the slope of the initial linear portion of the shear stress–shear strain curve.

The three-rail version of the test uses a different fixture and does not orient the specimen at an angle to the loading axis; see Figure 4.45. The shear strain is

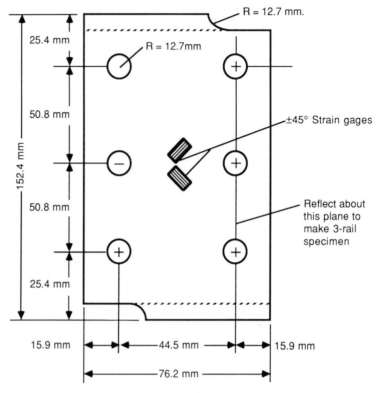

Figure 4.43 Rail shear specimen.

the same as for the two-rail test, but the shear stress is changed by a factor of one-half. The specimen is loaded in compression, essentially determining shear response in two specimens simultaneously. Although the more symmetric design of the three-rail test should produce better results, in practice, little difference is seen between results from the two methods.

The rail shear test method can be used to test unidirectional, cross-ply and angle-ply laminates, and discontinuous fiber composites for strength and modulus. The test does not work well on laminates with shear coupling (A^{16} and A_{26}) terms or where the Poisson's ratio approaches unity (i.e., ±45° laminates). The shear coupling allows the shear loading to produce significant tension and compression response in the material. For laminates a length-to-width aspect ratio for the gauge section of 10 has been suggested by Whitney (see [32]). Failures on unidirectional specimens often occur at the rails or in the gripped region as a result of bending stresses or slippage of the specimen and subsequent loading through bolt bearing. When this occurs, the strength data measured are not valid. Cross-ply laminates perform better.

Figure 4.44 Two-rail shear test fixture.

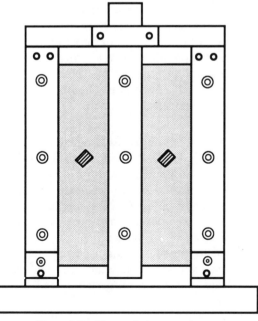

Figure 4.45 Three-rail shear test fixture.

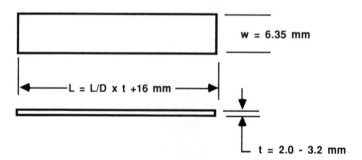

Figure 4.46 Short beam shear test specimen.

SHORT BEAM SHEAR TEST

The short beam shear test is a simple three-point bending test used to measure the apparent interlaminar shear strength of materials for quality control purposes. The specimen is designed to have a span-to-depth ratio of 4:1 and typically has the dimensions shown in Figure 4.46. It is an attractive specimen for quality control because of its small size and relative simplicity.

The test is performed as a typical flexure test. The shear strength is determined from the equation

$$\tau_{xz} = \frac{3P}{4bh} \qquad (34)$$

This calculation is based on the assumption that the material is homogeneous and loaded in pure beam bending, resulting in a parabolic shear stress distribution throughout the beam. This is not the case; only a very small portion of the beam is in a state of parabolic shear, and other portions of the beam have skewed shear stress distributions because of the concentrated load reactions under the loading noses. For this reason care should be exercised in confirming the failure mode of short beam shear specimens.

V-NOTCHED BEAM SHEAR

One of the newest and most promising shear test methods to be introduced is the V-notched beam shear test [33–37]. We perform the test by using either of two loading fixtures: the Iosipescu fixture shown in Figure 4.47 or the asymmetric four-point bending (AFPB) fixture shown schematically in Figure 4.48. Both fixtures use the same specimen, although tabs may be required on the AFPB specimen to prevent crushing under the load noses. Figure 4.49 describes the critical dimensions and geometry of the specimen.

The test can be used to measure strength and modulus of continuous or discontinuous reinforced composite materials. In unidirectional laminates, specimens can be fabricated to measure properties in the 1-2, 1-3, or 2-3 material planes (see Figure 4.50).

The test is conducted by mounting the specimen in the fixture and loading monotonically to failure. For modulus determination a strain rosette must be

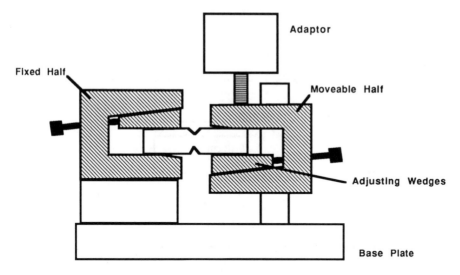

Figure 4.47   Iosipescu shear test fixture.

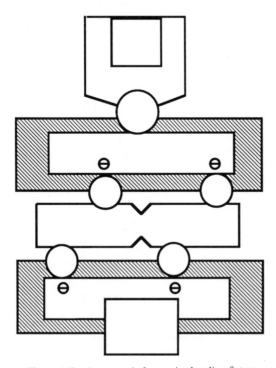

Figure 4.48   Asymmetric four-point bending fixture.

Without Tabs

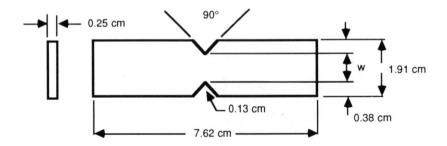

With Tabs

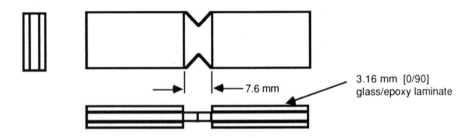

all other dimensions the same

Figure 4.49   V-notched shear test specimen.

mounted with the gauges oriented at $\pm 45°$ to the shear plane in the test section between the two notches. For the Iosipescu test the shear stress is determined by the equation

$$\tau_{12} = \frac{P}{wt} \tag{35}$$

The asymmetric four-point bending test uses a different equation for determining shear stress.

$$\tau_{12} = \frac{P^*}{wt} \tag{36}$$

where

$$P^* = \frac{P(a - b)}{(a + b)} = 0.4595$$

for $a = 0.5$ and $b = 1.35$ as specified for the test.

The shear strain for both methods is

$$\gamma_{12} = |\varepsilon_{45} - \varepsilon_{-45}| \tag{37}$$

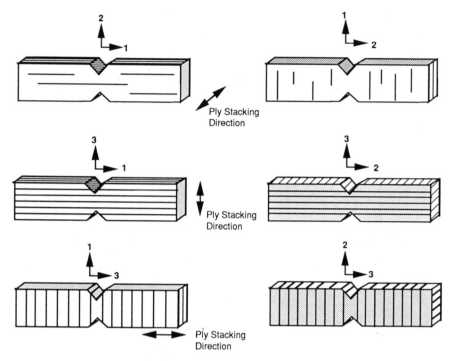

Figure 4.50 Specimen configurations for measuring shear properties in different planes.

The shear modulus is normally evaluated from the initial slope of the shear stress–strain curve.

Typical load deflection curves are shown for several specimen types in Figure 4.51. When a 0° unidirectional specimen for G_{12} is tested, the small load drop caused by the development of the horizontal crack at the notch root is ignored. Ultimate shear-stress is determined from shear failure in the test section. These results have been shown to correlate with strength results from $\pm 45°$ tension tests. Schematics of typical failures are shown in Figure 4.52.

The V-notched beam shear test has been thoroughly analyzed using elasticity and finite element methods for materials varying from isotropic to strongly orthotropic in properties [33, 37]. The geometry of the notch root has been chosen to minimize the stress concentration at the notch root and to maximize the percentage of the test section in uniform shear. While notch depth, notch tip radius, and angle all affect the stress in the test section, the notch root radius and quality are most critical. It is recommended that the notch be ground using a shaped grinding wheel or abrasive cutting wheel.

3.4.2 Summary Remarks on Shear Testing

When deciding which test method to use for determining shear properties of composite materials, many factors must be considered. The test method should be chosen to best fit the objectives of the characterization given the amount and

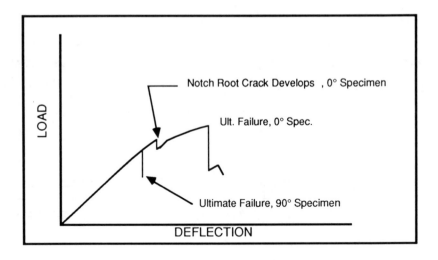

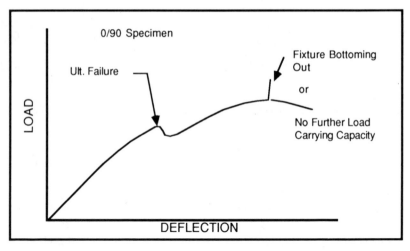

Figure 4.51 Typical load deflection curves.

form of material available for testing. Some of the test methods only allow measurement of strength, others, strength and modulus. Some methods only work for continuous fiber composites, while others work for both continuous and discontinuous fiber composites. Some require large specimens and precise machining, others, very small specimens simply machined. The accuracy of the information from the method reflects many of these factors. While torsion generates the purest state of stress, it is impractical for many composite forms. The ±45° tension test is one of the best for continuous fiber materials, while the V-notch beam shear test rivals the accuracy of the ±45° tension test; yet, it is more flexible, allowing the testing of discontinuous fiber composites in any principal material planes.

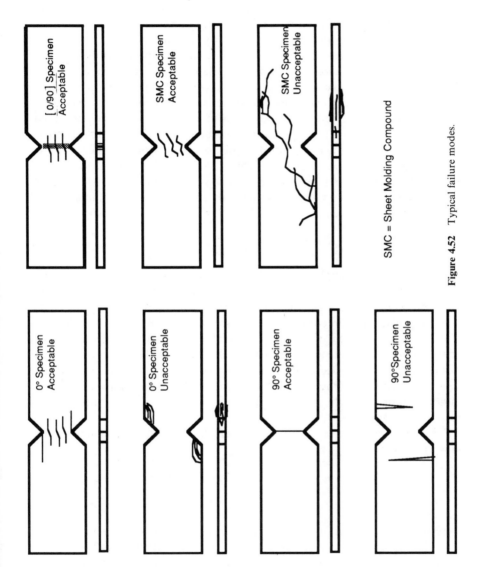

SMC = Sheet Molding Compound

Figure 4.52 Typical failure modes.

203

4   INTERLAMINAR FRACTURE OF COMPOSITES

Recently, with the introduction of composite materials into primary load bearing structures, the issue of damage tolerance is of increasing importance in design and materials selection. While damage tolerance testing is performed on substructure specimens or on prototype parts, the use of fracture tests has become popular for materials screening and qualification programs. Of particular interest is the characterization of interlaminar fracture properties that resist failure in the weakest material direction of the composite structure.

The short beam shear test is one of the methods most commonly applied to determine interlaminar properties, but it has many deficiencies as documented in the literature [38–41]. The fracture mechanics approach to interlaminar fracture characterization is a more fruitful approach. Fracture mechanics is more complex than the mechanical characterization methods described in Section 3. Instead of stress–strain we must look at strain energy release rates and fracture toughness. Given this jump in complexity, more fundamental background will be given in support of the description of these tests.

4.1   Fundamental Concepts of Fracture

The most commonly employed approach to characterize interlaminar fracture resistance of composites is by the strain energy release rate. The strain energy release rate G is a thermodynamic parameter that quantifies the energy available for crack growth. The fracture criterion based on G simply states that fracture (crack growth) will occur if the energy available for crack growth exceeds the energy required to create a unit of new crack area [42]. The energy approach does not require a detailed analysis of the stress state at the crack tip, where a stress singularity exists, thereby allowing the use of beam theory and strength of materials approaches to calculate the energy release rate of the fracture specimen [43, 44].

For a plate of thickness h containing a crack of length a, the condition for crack growth is

$$\frac{d(W - U)}{dA} = \frac{dS}{dA} \tag{38}$$

where W is the work performed by external forces, U is the elastic strain energy stored in the body, and S is the energy for crack formation [42–44]. The quantity $d(W - U)/dA$ is the net driving force for crack extension and is denoted by G. The term dA is the new crack area. The dS/dA term is the crack resistance. If the crack resistance is independent of crack length, this quantity is a material constant denoted by G_c, fracture toughness.

Crack extension can occur through three distinct fracture modes, as shown in Figure 4.53. Mode I is the crack opening mode, mode II is a shearing mode, and mode III is a tearing mode. A real failure process may occur as a result of any

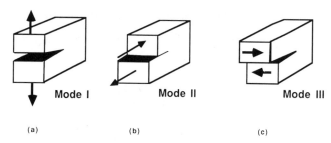

Figure 4.53 Modes of crack surface displacements: (a) mode I (opening), (b) mode II (forward shear), and (c) mode III (antiplane shear).

one or a combination of these modes [45–49]. In this chapter only mode I and mode II fracture test methods are covered. Mode III test methods are just beginning to be developed [50, 51].

Several tests, such as the edge delamination specimen, the Arcan test [52, 53], and the cracked lap shear specimen proposed by Wilkins (see [38]), measure mixed mode strain energy release rates and are not discussed in detail here. The double cantilever beam and end-notched-flex test methods will be the focus of these discussions.

4.2 Mode I Fracture Test—The Double Cantilever Beam Method

While it is not standardized, the double cantilever beam specimen for mode I fracture has been studied extensively by ASTM and is nearing standardization. First developed in a tapered form by Bascom (see [54]), the specimen has evolved to a straight-sided geometry as shown in Figure 4.54. The analysis developed for data reduction in the subsequent sections applies only to the straight-sided specimen.

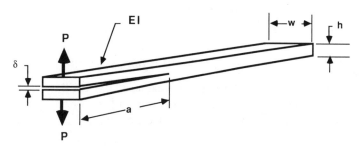

Figure 4.54 Double cantilever beam specimen shown with applied load P, crack length a, crack opening displacement δ, width w and thickness h. The bending stiffness of each beam half-thickness is EI.

4.2.1   Analysis of the Double Cantilever Beam Specimen

The strain energy release rate is defined in terms of specimen compliance C as follows:

$$G = \left(\frac{P^2}{2w}\right)\left(\frac{dC}{dA}\right) \tag{39}$$

The compliance of the DCB specimen as obtained from beam theory is

$$C = \frac{2a^3}{3E_1 I} \tag{40}$$

where a is the crack length and $E_1 I$ is the flexural rigidity of the specimen (see Figure 4.54).

The strain energy release rate G_I (subscript "I" for mode I) can be obtained by differentiating (40) with respect to crack length and substituting into (39) to get

$$G_I = \frac{P^2 a^2}{w E_1 I} \tag{41}$$

The onset of crack growth occurs when G_I reaches its critical value G_{Ic}. In terms of load this corresponds to $P = P_c$. Consequently,

$$G_{Ic} = \frac{P_c^2 a^2}{w E_1 I} \tag{42}$$

It should be recognized that (42) is only an approximation to the strain energy release rate. Discrepancies may occur if the beams also undergo significant shear deformation. For beams with large Young's modulus-to-interlaminar shear modulus ratio (E_1/G_{13}) and of large thickness-to-crack length (h/a), shear deformation is significant and (42) will be in error. Another potential source of error is the nonlinear deformation resulting from large deflections of the beams at long crack lengths [55]. This can be minimized by proper specimen design.

4.2.2   Sizing the Double Cantilever Beam Specimen

Interlaminar shear deformation in the beams of the DCB specimen can be minimized by considering the following expression [56] for the deflection at the point of load application of a cantilever beam of length L and thickness t.

$$\Delta = \Delta_B \left[1 + 0.3 \left(\frac{t}{L}\right)^2 \frac{E_1}{G_{13}} \right] \tag{43}$$

where $\Delta_B = PL^3/(3E_1 I)$ is the beam deflection neglecting interlaminar shear deformation. For the DCB, $t = h/2$ and $L = a$, giving

$$\Delta = \Delta_B \left[1 + \left(\frac{3h^2}{40a^2} \right) \left(\frac{E_1}{G_{13}} \right) \right] \tag{44}$$

By this equation shear deformation can be minimized to a certain fraction f_s, where $f_s = 3h^2 (E_1/G_{13})/(40a^2)$. Assuming $f_s \leqslant 5\%$, we get

$$\frac{a}{h} \geqslant \sqrt{\frac{3(E_1/G_{13})}{2}} \tag{45}$$

For a typical graphite/epoxy material E_1/G_{13} is approximately 25 [44] and $a/h \geqslant 6.1$. This means that for a 24-ply laminate the minimum crack length should be at least 20 mm. An upper boundary on crack length has been derived in [57] based on the requirement of small deflection linear elastic load displacement response. This yields

$$a \leqslant 9.2 \times 10^{-2} \sqrt{\frac{E_1 h^3}{G_{Ic}}} \tag{46}$$

For the same graphite/epoxy beam, $E_1 = 140 \, \text{GPa}$ and $G_{Ic} = 0.19 \, \text{kJ/m}^2$, the maximum permissible crack length is 42 cm, well beyond that required for most specimens.

4.2.3   Specimen Fabrication

A suggested panel configuration for the fabrication of DCB specimens is shown in Figure 4.55. A starter crack is created by inserting a folded Teflon or Kapton film at the laminate center as shown in the figure. The film prevents bonding between the plies during processing of the laminate. A thin, 0.025-mm film is suggested to minimize thickness increase and to create a sharp crack. The procedure described previously should be used to size specimens, but for carbon fiber materials a 24-ply laminate is commonly used. This may require modification for very tough resin systems, but it works reasonably well for most systems currently used. The specimens should be cut to size with a diamond wafing saw.

The hinged specimen is preferred and should be bonded to the specimen according to Figure 4.56 by using proper surface preparation and a suitable strength adhesive. Alignment of the hinges is critical as is frictionless rotation of the hinge. The edges of the specimen should be painted white to enhance visualization of the crack tip during testing.

4.2.4   Double Cantilever Beam Testing

The test specimens must be mounted and tested in a properly aligned and calibrated test frame. The crack length, as defined in Figure 4.54, should be

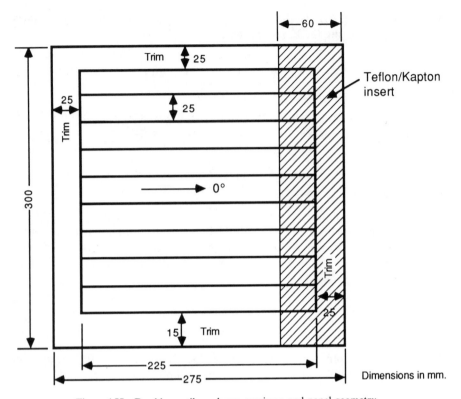

Figure 4.55   Double cantilever beam specimen and panel geometry.

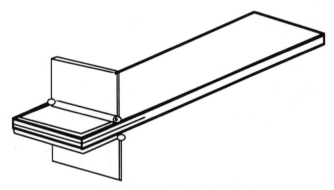

Figure 4.56   Hinged DCB specimen.

measured with a precision dial caliper on both sides of the specimen. Crack length a is defined as the distance between the line of load application and the crack tip. The load-point displacement δ is defined as the displacement of the points of load application, Figure 4.54; and we determine it most accurately by using an extensometer or linear variable differential transformer, but it can be determined adequately using crosshead displacement corrected for machine compliance. The test should be performed at a crosshead speed of about 2 mm/min, and the load versus displacement curve should be plotted on the chart recorder. The specimen should be loaded until the crack extends a short distance and then the crosshead stopped. These data are usually not reliable and are ignored; the procedure serves to precrack the specimen and to create a natural crack front. After the new crack length is measured, the specimen is reloaded until the crack grows a short distance further. One of two procedures can be used to proceed. The crosshead can be stopped, and the specimen unloaded slightly; the new crack length can be measured and the procedure repeated. Alternatively, the specimen can be premarked at the desired crack extension intervals and the load marked (pipped) as the crack tip reaches the mark for each crack length. Either procedure should be carried out until the crack has extended to approximately 15 cm. The results will be a curve similar to that depicted in Figure 4.57. Fibers may bridge the crack surfaces at longer crack lengths causing increased resistance to crack propagation and unrealistically high fracture toughness values.

4.2.5 Double Cantilever Beam Data Reduction

The critical strain energy release rate G_{Ic} is calculated from the compliance and critical load versus crack length data collected during the test. The compliance is established for each crack length from the inverse for the slope of the curve P versus δ, shown in Figure 4.57

$$C = \delta/P \qquad (47)$$

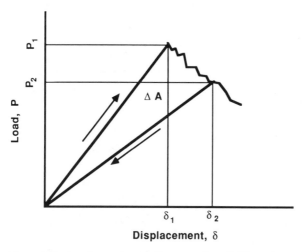

Figure 4.57 Loading and unloading curve for a DCB specimen.

From (40) (Section 4.2.1) the compliance can be expressed as

$$C = A_1 a^3 \qquad (48)$$

where $A_1 = 2/(3E_1 I)$.

To determine A_1, a plot of the log C versus log a yields

$$\log C = \log A_1 + 3 \log a \qquad (49)$$

that is, a straight line of slope 3 with intercept $\log A_1$ on the y axis at $\log a = 0$; see Figure 4.58. Once it is known, A_1 can be used in the following equation to determine the strain energy release rate at any crack length.

$$G_I = \frac{3A_1 a^2}{2w} \qquad (50)$$

To determine the critical strain energy release rate we need to determine the critical load P_c for the onset of crack growth. From (42)

$$P_c = \frac{\sqrt{wE_1 IG_{Ic}}}{a} \qquad (51)$$

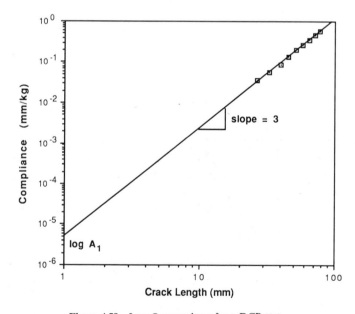

Figure 4.58 Log C versus log a for a DCB test.

Assuming that G_{Ic} is a constant,

$$P_c = \frac{A_2}{a} \tag{52}$$

where

$$A_2 = \sqrt{wE_1IG_{Ic}} \tag{53}$$

To determine the constant A_2, a plot of the $\log P_c$ versus $\log a$ should yield

$$\log P_c = \log A_2 - \log a \tag{54}$$

a straight line with a slope of -1 and an intercept $\log A_2$ at $\log a = 0$; see Figure 4.59. Using A_1 and A_2 we can determine the critical strain energy release rate using

$$G_{Ic} = \frac{3A_1A_2^2}{2w} \tag{55}$$

A second method, available to reduce the data obtained in the DCB test, is termed the area method. The critical strain energy release rate may be

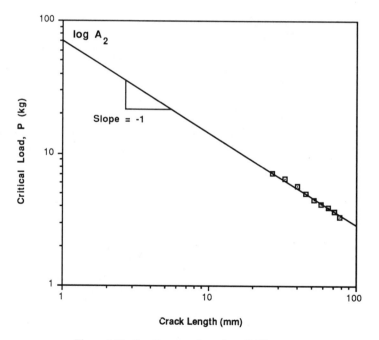

Figure 4.59   Log P_c versus log a for a DCB test.

determined from a loading–unloading sequence according to Figure 4.57 as follows:

$$G_{Ic} = \frac{\Delta A}{w \Delta a} \qquad (56)$$

where ΔA is the area indicated in Figure 4.57 and Δa is the increment in crack length. This is approximated by

$$G_{Ic} = \frac{P_1 \delta_2 - P_2 \delta_1}{2 w \Delta a} \qquad (57)$$

An average G_{Ic} value is obtained from the total series of loading and unloading curves. The two methods yield similar results. The area method yields an average toughness value for propagation of the crack, while the compliance method yields only the toughness for onset of crack growth.

4.3   Mode II Fracture Testing—End-Notched-Flex Test Method

The end-notched-flex (ENF) test method introduced by Russell and Street [49, 58] is one of the most promising methods currently available for mode II shear fracture, and like the DCB method it is currently being studied intensely by ASTM for potential standardization. The method was analyzed in detail by Gillespie and co-workers [59], and the purity of the mode II fracture mechanism has been verified.

Figure 4.60 shows the geometry of the ENF specimen, which is loaded in three-point flexure to produce the shear stresses necessary to propagate the crack embedded at the laminate midsurface. The test is used most commonly with unidirectional laminates, but it is not restricted to this case. The geometric and loading interactions for the ENF specimen are complex. The specimen is asymmetric—the crack is placed at one end—and friction between the crack surfaces can influence results [59–61] if it is not minimized by suitable specimen design. As with most bending configurations, transverse shear deformation must also be considered.

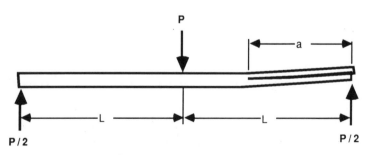

Figure 4.60   End-notched-flex specimen shown with span $2L$, applied load P, and crack length a.

4.3.1   Compliance and Strain Energy Release Rate for the End-Notched-Flex Specimen

The compliance for the ENF specimen incorporating transverse shear deformation [59] is

$$C = \frac{2L^3 + 3a^3}{8E_1 w h^3}\left[1 + \frac{2(1.2L + 0.9a)h^2 E_1}{(2L^3 + 3a^3)G_{13}}\right] \tag{58}$$

The geometry symbols in this equation are partly shown in Figure 4.60. The total span is $2L$, total thickness $2h$, crack length a, and width w. The material properties E_1 and G_{13} are the Young's modulus and interlaminar shear modulus, respectively. For materials with large shear modulus and small thickness-to-crack length ratio (h/a), the second term inside the brackets of (58) becomes negligible, and classical beam theory is recovered.

Using the compliance expression the strain energy release rate G_{II} becomes

$$G_{II} = \frac{9a^2 P^2}{16E_1 w^2 h^3}\left[1 + 0.2\left(\frac{E_1}{G_{13}}\right)\left(\frac{h}{a}\right)^2\right] \tag{59}$$

which reduces to

$$G_{II} = \frac{9a^2 P^2}{16E_1 w^2 h^3} \tag{60}$$

for cases where classical beam theory applies.

Friction between the sliding crack surfaces may dissipate energy that can contribute to the apparent fracture energy recorded in the ENF test. Using a Griffith energy balance, we can define a nondimensional energy release rate parameter $g(\mu)$ [59] as

$$g(\mu) = \frac{G_{II}(\mu = 0) - G_{II}(\mu)}{G_{II}^{BT}} \tag{61}$$

where μ is the coefficient of sliding friction. The parameter $g(\mu)$ quantifies the relative reduction in energy release rate because of friction between the crack surfaces, and substitution of quantities found in [59] into (61) yields a simple expression

$$g(\mu) = \frac{4\mu h}{3a} \tag{62}$$

which will be used in specimen design to minimize friction.

4.3.2   Specimen Design Considerations

The ENF specimen must be designed such that fracture occurs prior to the onset of material or geometrical nonlinearities. These requirements lead to a minimum thickness of the beam. The upper bound on the beam thickness is controlled by frictional considerations.

The requirement for small deflection behavior yields the minimum semithickness h_{min}

$$h_{min} = \left[\frac{G_{IIc}(L^2 + 3a^2)^2}{4(v_a')^2 a^2 E_1} \right]^{1/3} \tag{63}$$

where G_{IIc} is the mode II fracture toughness and v_a' is the maximum allowable slope of the deflection curve ($v_a' = 0.2$) [60].

The requirement that the maximum bending stress at any point within the ENF specimen should not exceed the yield point or the ultimate stress in order to avoid material nonlinearities yields

$$h_{min} = \frac{L^2 G_{IIc}}{a^2 \varepsilon_a^2 E_1} \tag{64}$$

where ε_a is the allowable strain. The minimum thickness is dictated by the condition that applied, which is the condition that requires the largest thickness.

To determine the maximum allowable thickness or a minimum crack length to minimize the influence of friction it is required that $g(\mu)$ be small, say 2–3%. Given this for a fixed thickness, nominally $h = 1.7$ mm, the minimum crack length is 17 mm, well below the specified length of 25 mm.

4.3.3 Fabrication of the End-Notched-Flex Specimen

Fabrication of the ENF specimen is very similar to that for the DCB specimen. A recommended panel layout is shown in Figure 4.61. One major difference between the DCB and ENF specimens is that a precrack must be generated for the ENF specimen prior to testing since crack growth is unstable. One method of producing the crack is to clamp the specimen at a location ahead of the implanted crack front and use a razor blade wedge to grow the crack in mode I fracture. The mode I starter crack produces conservative G_{IIc} values [62]. For ductile materials a mode II precracking procedure may be used to achieve more linear load displacement curves. To enhance detection of the initial crack length the specimen edges should be painted with a white flat enamel.

4.3.4 End-Notched-Flex Testing

The specimen is mounted in a three-point bending fixture such that the total span ($2L$) is 100 mm with the crack length equal to 25 mm (see Figure 4.60). This means that the distance from the outer load support to the crack front is 25 mm. The specimen is then loaded with a crosshead speed of 5 mm/min, and the displacement is measured under the central loading nose with a deflectometer. Again the crosshead travel can be used if correction is made for machine compliance. A real time plot of load versus displacement such as that shown in

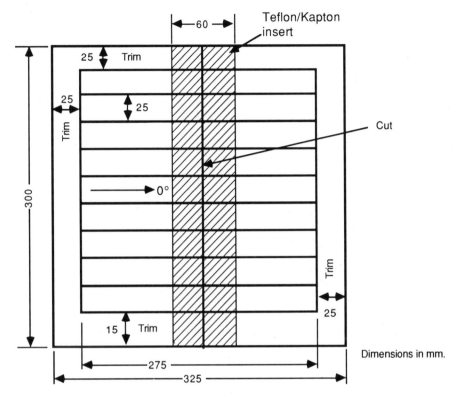

Figure 4.61   End-notched-flex specimen and panel geometry and dimensions.

Figure 4.62 should be recorded for each specimen. A more accurate check of initial crack length can be made by cracking the failed specimen into two parts and measuring the distance between the mode I crack front and the indentation on the specimen surface from the outer load nose. In the case of mode II precracking, the crack length is based on edge observations. A slight variation of this method is discussed in Section 4.3.5 for the compliance calibration data reduction scheme.

4.3.5   End-Notched-Flex Data Reduction

Two methods for reducing G_{IIc} from load versus displacement records of ENF tests are described. The first method is a direct method based on (59). The alternative method is to determine the compliance for several crack lengths and numerically establish the slope dC/da required for G_{IIc} determination according to (39). This method, while requiring more time, may be more accurate since some of the beam theory assumptions are removed. Both methods assume linear elastic response up to failure as shown in Figure 4.62. For very tough or ductile

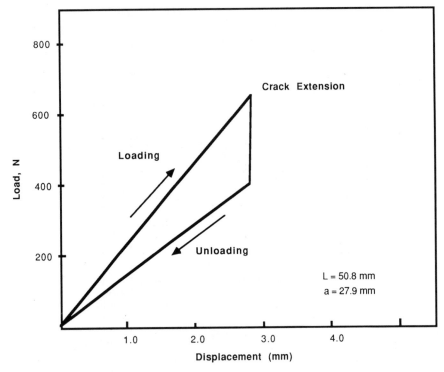

Figure 4.62 Typical load-deflection curve for a 25.4-mm-wide ENF specimen made from AS4/3501-6 graphite/epoxy. The half-span L is 50.8 mm and initial crack length a is 27.9 mm.

polymers, like PEEK, nonlinearities can occur and require special attention [62].

From the load-displacement curve shown in Figure 4.62 the compliance is found from the inverse of the slope $dP/d\delta$. The toughness is then determined using the critical load P_c in the following equation:

$$G_{IIc} = \frac{9a^2 P_c^2 (C - C_{SH})}{4wL^3 [1 + 1.5(a/L)^3]} \tag{65}$$

where

$$C_{SH} = \frac{6L + 3a - \left(\dfrac{L^3}{a^2}\right)}{20whG_{13}} \tag{66}$$

The inplane shear modulus G_{12} can be used as an approximation to the interlaminar shear modulus G_{13}.

The alternate method to determine G_{IIc} is to perform compliance calibration. This method requires a long ENF specimen with a long precrack so that various crack lengths can be achieved by sliding the specimen across the three-point bending fixture. The specimen is loaded below the critical load to avoid any crack extension, and the load deflection curves are plotted for the different crack lengths. After completion of the compliance measurements, fracture testing is performed. A third-order polynomial is fitted to the data by linear regression

$$C = C_1 + ma^3 \qquad (67)$$

where C_1 is a constant and m is the slope of the line C versus a^3. Differentiation of (67) and substitution into (39) yields

$$G_{IIc} = \frac{3P_c^2 ma^2}{2w} \qquad (68)$$

where P_c is the critical load for crack extension at the crack length a for which the testing was performed. The two methods produce similar results for materials where beam theory assumptions hold [63–65]. Cases where beam theory assumptions do not hold, like laminates with variable volume fraction through the thickness, produce significant differences in results between the two methods with the compliance calibration procedure being more accurate [66].

4.4   Summary

In this chapter we discussed methods for the measurement of simple mode I and mode II fracture properties in composite materials. These tests are useful in the evaluation of matrix toughness and interfacial bonding between the fiber and matrix phases. The quantitative values of strain energy release rate or fracture toughness do not have direct application to design, but they provide a basis for developing property comparisons. When making property comparisons between different material systems, it is important to perform fractographic evaluations of failed specimens to verify failure mechanisms and assess contributions because of fiber bridging or fiber fracture.

References

1.   R. Jones, *Mechanics of Composite Materials*, Scripta Book Company, Washington, DC, 1975.
2.   S. Tsai and H. Hahn, *Introduction to Composite Materials*, Technomic Publishing Company, Westport, CT, 1980.
3.   D. W. Wilson, R. B. Pipes, and R. L. McCullough, Eds., *University of Delaware Composites Design Encyclopedia*, Center for Composite Materials, University of Delaware, Newark, DE, 1986.

4. J. R. Vinson and T-W. Chou, *Composite Materials and Their Use in Structures*, Halstead, Applied Science Publishers, Barking, Essex, England, 1975.

5. ASTM Standard D3800, *Standard Test Method for Density of High-Modulus Fibers*, *Annual Book of Standards*, Vol. 15.05, American Society for Testing and Materials, Philadelphia, PA, 1989 (or most current printing).

6. ASTM Standard D1505, *Standard Test Method for Density of Plastics by the Density-Gradient Technique*, *Annual Book of Standards*, Vol. 8.01, American Society for Testing and Materials, Philadelphia, PA, 1989 (or most current printing).

7. ASTM Standard D2584, *Standard Test Method for Ignition Loss of Cured Reinforced Resins*, *Annual Book of Standards*, Vol. 15.05, American Society for Testing and Materials, Philadelphia, PA, 1989 (or most current printing).

8. ASTM Standard D3171, *Standard Test Method for Fiber Content of Resin-Matrix Composites By Matrix Digestion*, *Annual Book of Standards*, Vol. 15.05, American Society for Testing and Materials, Philadelphia, PA, 1989 (or most current printing).

9. ASTM Standard D2734, *Standard Test Method for Void Content of Reinforced Plastics*, *Annual Book of Standards*, Vol. 15.03, American Society for Testing and Materials, Philadelphia, PA, 1989 (or most current printing).

10. ASTM Standard D3039, *Standard Test Method for Tensile Properties of Fiber-Resin Composites*, *Annual Book of Standards*, Vol. 15.03, American Society for Testing and Materials, Philadelphia, PA, 1989 (or most current printing).

11. ASTM Standard D638, *Standard Test Method for Tensile Properties of Plastics*, *Annual Book of Standards*, Vol. 8.01, American Society for Testing and Materials, Philadelphia, PA, 1989 (or most current printing).

12. D. E. Walrath and D. F. Adams, *Exp. Mech.*, **23(1)**, 105 (1983).

13. D. E. Walrath and D. F. Adams, *Verification and Application of the Iosipescu Shear Test Method*, UWME-DR-401-103-1, University of Wyoming, Laramie, WY, 1984.

14. M. G. Abdallah, D. S. Gardner, and H. E. Gascoigne, "An Evaluation of Graphite/Epoxy Iosipescu Shear Specimen Testing Methods With Optical Techniques," *Proceedings of the 1985 SEM Spring Conference on Experimental Mechanics*, Society of Experimental Mechanics, Philadelphia, PA, 1985, p. 833.

15. ASTM Standard 3379, *Test Method for Tensile Strength and Youngs Modulus for High-Modulus Single-Filament Materials*, *Annual Book of Standards*, Vol. 15.05, American Society for Testing Materials, Philadelphia, PA, 1989 (or most current printing).

16. ASTM Standard 4018, *Test Method for Tensile Properties of Continuous Filament Carbon and Graphite Yarns, Strands, Rovings and Tows*, *Annual Book of Standards*, Vol. 15.05, American Society for Testing and Materials, Philadelphia, PA, 1989 (or most current printing).

17. D. W. Oplinger, K. R. Ganhi, and B. S. Parker, *Studies of Tension Test Specimens for Composite Material Testing*, AMMRC TR 82-27, Army Materials and Mechanics Research Center, Watertown, MA, 1982.

18. ASTM Standard D695, *Standard Test Method for Density of High-Modulus Fibers*, *Annual Book of Standards*, Vol. 15.05, American Society for Testing and Materials, Philadelphia, PA, 1989 (or most current printing).

19. ASTM Standard D3410, *Standard Test Method for Compressive Properties of Unidirectional or Crossply Fiber-Resin Composites*, *Annual Book of Standards*, Vol.

15.05, American Society for Testing and Materials, Philadelphia, PA, 1989 (or most current printing).

20. M. J. Shuart, *An Evaluation of the Sandwich Beam Compression Test Method for Composites, Test Methods and Design Allowables for Fibrous Composites*, ASTM STP 734, American Society for Testing and Materials, Philadelphia, PA, 1981, p. 152.

21. Y. M. Tarnopol'skii and T. Kincis, *Static Test Methods for Composites*, Van Nostrand Reinhold, New York, 1985.

22. J. M. Whitney, I. M. Daniel, and R. B. Pipes, *Experimental Mechanics of Fiber Reinforced Composite Materials*, Society for Experimental Stress Analysis, Brookfield Center, CT, 1982.

23. J. M. Whitney, C. E. Browning, and A. Mair, "Analysis of the Flexure Test for Laminated Composite Materials," *Composite Materials: Testing and Design (Third Conference)*, ASTM STP 546, American Society for Testing and Materials, Philadelphia, PA, 1974, p. 30.

24. C. Zweben, W. S. Smith, and M. W. Wardle, "Test Method for Fiber Tensile Strength, Composite Flexural Modulus and Properties of Fabric Reinforced Laminates," *Composite Materials: Testing and Design (Fifth Conference)*, ASTM STP 674, American Society for Testing and Materials, Philadelphia, PA, 1979, p. 228.

25. ASTM Standard D3518, *Inplane Shear Stress-Strain Response of Unidirectional Reinforced Plastics, Annual Book of Standards*, Vol. 15.05, American Society for Testing and Materials, Philadelphia, PA, 1989 (or most current printing).

26. ASTM Standard D4255, *Inplane Shear Properties of Composite Laminates, Annual Book of Standards*, Vol. 15.05, American Society for Testing and Materials, Philadelphia, PA, 1989 (or most current printing).

27. ASTM Standard D3846, *Inplane Shear Strength of Reinforced Plastics, Annual Book of Standards*, Vol. 8.01, American Society for Testing and Materials, Philadelphia, PA, 1989 (or most current printing).

28. S. Lee and M. Munro, *Composites*, **17(1)**, 13 (1986).

29. J. M. Whitney, I. M. Daniel, and R. B. Pipes, *Experimental Mechanics of Fiber Reinforced Composite Materials*, Society for Experimental Stress Analysis, Brookfield Center, CT, 1982, p. 189.

30. I. M. Daniel, *Biaxial Testing of Graphite/Epoxy Components Containing Stress Concentrations, Air Force Technical Report AFML-TR-76-274*, Part I, Wright Patterson Air Force Base, OH, 1968.

31. N. J. Pagano and J. C. Halpin, *J. Compos. Mater.*, **2(1)**, 18 (1968).

32. J. M. Whitney, D. L. Stansbarger, and H. B. Howell, *J. Compos. Mater.*, **5(1)**, 24 (1971).

33. B. S. Spigel, "An Experimental and Analytical Investigation of the Iosipescu Shear Test for Composite Materials," unpublished masters thesis, Old Dominion University, Norfolk, VA, 1984.

34. D. F. Adams and D. E. Walrath, *J. Compos. Mater.*, **21**, 494 (1987).

35. D. F. Adams and D. E. Walrath, *Exp. Mech.*, **27(2)**, 113 (1987).

36. D. F. Adams and D. E. Walrath, *J. Compos. Mater.*, **21**, 494 (1987).

37. P. Ifju and D. Post, "A Compact Double-Notched Specimen for In-Plane Shear

Testing," *Proceedings of the 1989 SEM Spring Conference on Experimental Mechanics*, Society for Experimental Mechanics, Philadelphia, PA, May, 1989.

38. D. J. Wilkins, J. R. Eisenmann, R. A. Camin, W. S. Margolis, and R. A. Benson, "Characterizing Delamination Growth in Graphite-Epoxy," *Damage in Composite Materials*, ASTM STP 775 91982, American Society for Testing and Materials, Philadelphia, PA, 1989, p. 168.

39. ASTM Standard D2344, *Apparent Interlaminar Shear Strength of Composite Materials, Annual Book of Standards*, Vol. 15.05, American Society for Testing and Materials, Philadelphia, PA, 1989 (or most current printing).

40. J. M. Whitney and C. E. Browning, *Exp. Mech.*, **25(3)**, 294 (1985).

41. J. M. Whitney, *Compos. Sci. Tech.*, **22**, 167 (1985).

42. D. Broek, *Elementary Engineering Fracture Mechanics*, 3rd ed., Martinus-Nijhoff, Alphen aan den Rijn, The Netherlands, 1984.

43. J. M. Whitney, I. M. Daniel, and R. B. Pipes, *Experimental Mechanics of Fiber Reinforced Composite Materials*, rev. ed., Prentice-Hall, Englewood Cliffs, NJ, 1987.

44. L. A. Carlsson and R. B. Pipes, *Experimental Characterization of Advanced Composite Materials*, Prentice-Hall, Englewood Cliffs, NJ, 1987.

45. J. M. Whitney, C. E. Browning, and W. Hoogsteden, *J. Reinf. Plast. Compos.*, **1**, 297 (1982).

46. A. S. D. Wang and F. W. Crossman, *J. Compos. Mater. Suppl.*, **14**, 71 (1980).

47. J. D. Whitcomb, *J. Compos. Mater.*, **15**, 403 (1981).

48. S. S. Wang, *J. Compos. Mater.*, **17**, 210 (1983).

49. A. J. Russell and K. N. Street, "Factors Affecting the Interlaminar Fracture Energy of Graphite/Epoxy Laminates," in T. Hayashi, K. Kawata, and S. Umekawa, Eds., *Progress in Science and Engineering Composites*, ICCM-IV, ASM International, Tokyo, 1982, p. 279.

50. S. L. Donaldson, "Mode III Interlaminar Fracture Characterization of Composite Materials," unpublished masters thesis, University of Dayton, Dayton, OH, 1987.

51. G. J. Becht, "Interlaminar Fracture of Composite Materials Under Mode III Loading," unpublished masters thesis, University of Delaware, Newark, DL, 1988.

52. M. Arcan, Z. Hashin, and A. Voloshin, *Exp. Mech.*, **18**, 141 (1978).

53. R. A. Jurf and R. P. Pipes, *J. Compos. Mater.*, **16**, 386 (1982).

54. W. B. Bascom, R. J. Bitner, R. J. Moulton, and A. R. Siebert, *Composites*, **11**, 9 (1980).

55. D. F. Devitt, R. A. Shapery, and W. L. Bradley, *J. Compos. Mater.*, **14**, 270 (1980).

56. S. P. Timoshenko, *Strength of Materials*, Part 1, Kreiger, Melbourne, FL, 1984.

57. J. W. Gillespie, Jr., L. A. Carlsson, R. B. Pipes, R. Rothchilds, B. Trethewey, and A. Smiley, *NASA Contract Rep.*, NAB-1-475 (1985).

58. A. J. Russell and K. N. Street, "Moisture and Temperature Effects on the Mixed-Mode Delamination Fracture of Unidirectional Graphite/Epoxy," *Delamination and Debonding of Materials*, ASTM STP 876, American Society for Testing and Materials, Philadelphia, PA, 1985, p. 349.

59. J. W. Gillespie, Jr., L. A. Carlsson, and R. B. Pipes, *Compos. Sci. Tech.*, **26**, 177 (1986).

60. L. A. Carlsson, J. W. Gillespie, Jr., and R. B. Pipes, *J. Compos. Mater.*, **20**, 594 (1986).

61. S. Mall and K. N. Kochnar, *J. Compos. Tech. Res.*, **8**, 54 (1986).

62. L. A. Carlsson, J. W. Gillespie, Jr., and B. R. Trethewey, *J. Reinf. Plast. Compos.*, **5**, 170 (1986).

63. L. A. Carlsson and J. W. Gillespie, Jr., "Mode II and Mixed Mode Interlaminar Fracture of Composites," in K. Friedrich, Ed., *Application of Fracture Mechanics to Composite Materials*, Elsevier, New York, 1989, p. 113.

64. J. D. Barrett and R. O. Foshi, *Eng. Fract. Mech.*, **9**, 371 (1977).

65. L. A. Carlsson, J. W. Gillespie, Jr., and B. R. Trethewey, "Finite Element and Plate Theory Based Design and Data Reduction for the ENF Fracture Specimen," *Proceedings From the Second Conference on Composite Materials*, Sponsored by the American Society for Composites and the University of Delaware, Technomic Publishing Company, Westport, CT, 1987, p. 399.

66. T. K. O'Brien, G. B. Murri, and S. A. Salpekar, *NASA Tech. Memo.*, 89157 (1987).

Chapter **5**

MECHANICAL TESTING
OF ADHESIVES

K. Lawrence DeVries, Carol R. Johnsen, and
Garron P. Anderson†

1   INTRODUCTION

The fundamental mechanism that determines how one material adheres to another material has never been clearly identified. Indeed, it appears that different mechanisms may be active in different systems. Despite extensive, careful research no definite, universally accepted relationship has been established between specific atomic or molecular parameters at or near an interface and the strength of an adhesive bond. While the purpose of this chapter is to explore the measurement of mechanical properties of joints, a brief outline of a few proposed theories for adhesion may be enlightening and may provide insight into the interpretation of physical test results. For details and specifics about these models or theories, the reader is referred to [1–4]. A great many chemical theories and phenomena have been proposed to explain adhesion. Only a few will be described briefly in this chapter. The following theories are perhaps those most widely cited and used to explain adhesion.

†Deceased, October, 1988.

Mechanical interlocking is different from chemical adhesion; nevertheless, this physical phenomena is likely involved in the mechanical strength of some practical joints. Surface roughening and some surface modification treatments serve the purpose of improving mechanical interlocking or *hooking*. For a graphic example of one way that mechanical interlocking can be used to enhance adhesion, the reader is referred to a model of aluminum that has been anodized to enhance adhesion, as proposed by Venables and co-workers [5]. This model demonstrates how effective anodizing develops a coherent, tightly attached, open meshlike oxide layer through which the adhesive can flow, thereby forming strong mechanical interlocks.

Diffusion theory deals with the hypothesis that one material interdiffuses into and with another. A primary advocate of this theory is Voyutskii [6]. This theory lends itself most readily to polymer bonding, where it is envisioned that a boundary layer is developed, along which the polymer chains of the two materials are intertwined. For this to occur, at least one of the polymers must exhibit some solubility in the other. The fact that viscosity, polymer type, temperature, and bonding times play an important role in determining joint strength is cited as evidence in support of the theory. It appears reasonable that the diffusion mechanism is likely to be involved in joining operations such as solvent bonding, which is commonly used to bond poly(methyl methacrylate) (PMMA) to PMMA, poly(vinyl chloride) (PVC) to PVC, and acrylonitrile butadiene styrene (ABS) to ABS. The fact that adhesion occurs between materials where this does not appear to be the case can be viewed as an argument against the general applicability of the theory. Campion [7] presented other structural arguments against the general use of this theory.

Absorption theory involving secondary molecular forces has also been proposed as a mechanism for adhesion [8–12]. This theory hypothesizes that molecules near the interface are attracted to each other by London dispersion forces, dipole–dipole interactions, hydrogen bonding, or other secondary molecular forces. The strength of these forces varies from 0.1 to 10 J/mol. Almost all adhesives are known to exhibit at least some dipole interactions, but it is difficult to account for the strength of many practical joints purely on the basis of such forces.

The *chemical reaction theory* proposes that chemical reactions occur between the adhesive and the adherent, such that primary chemical bonds are formed [9, 13–15]. While it is unlikely that this theory is universally applicable to adhesives, chemical reactions may well be present in some cases. For example, silanes [16–18] are sometimes used as coupling agents. These are bifunctional molecules in which one *end* is intended to interact with a polymeric adhesive while the other end reacts chemically with atoms on the adherend's surface layer, such as the oxygen in an oxide layer of a metal or the oxygen in a ceramic.

The *electrostatic force model of adhesion* assumes that the electrons within the adhesive and the adherend occupy different energy levels and electron transfer occurs across the surface [19, 20]. The surfaces are attracted to each other as a result of these opposite charges. It is generally assumed that these forces are not the primary contributors to the strength of most practical bonds.

The *acid–base reaction theory of adhesion* has been proposed [21, 22] to explain several observed adhesive phenomena. This theory is based on acid–base reactions at the surface. Initially, only the Brønsted concept of acid–base reactions was considered, but the current, more general theory incorporates the Lewis acid concept (electron donor–acceptor). The determination of the acidity or basicity of polymers is not as straightforward as may be expected. Several approaches that have been used for this determination are described in [22], as well as evidence for the theory's applicability to explain adhesive phenomena.

Adhesion is a system phenomena. The strength of a joint may, and for most practical joints does, involve many different factors. Crack growth in general and adhesive crack growth in particular can involve events that are somewhat removed from the crack front. It has been observed that joint strength depends on many factors such as adhesive and adherend thickness, details of joint and gross specimen geometry, loading rate, temperature, and other factors. Furthermore, there is seldom, if ever, a plane of demarcation or sharp interface in practical joints because the adhesives generally form more or less diffuse regions at the boundary between the adherends that cannot be realistically represented by a plane. This led to the suggested use of the term *interphase* rather than *interface* [23]. The strength of an adhesive joint is affected not only by the structure and details of this interphase region, but also by the structure and details well removed from it.

Other than *strength tests*, which are discussed later in Section 3, the most commonly conducted adhesive test is probably the measurement of contact angle [24–31]. These experiments provide a quantitative measurement of the wetting of a material by a liquid. It is important to note that the work of adhesion, as determined by wetting experiments, represents only a small portion of the total adhesive energy, which is described later. Nevertheless, it is generally essential that an adhesive, during some phase of its application, thoroughly wets the adherend in order to form the intimate contact necessary for strong bonding. Depending on the adhesive system used, this can occur in the melt for hot melt adhesives, before cure for epoxies and other two-part adhesives, before evaporation and absorption into the substrate of the solvent or carrier liquid for solvent or emulsion adhesives, and so on. Determination of the contact angle is the most common method used to measure the tendency of a liquid to *wet* a surface. Thorough discussions of the Young–Dupre equation and ways to measure and interpret the contact angle between a drop and a surface are found in many textbooks on surface chemistry and adhesion [1, 24, 25].

The fact that it has not, in general, been possible to unambiguously identify the exact mechanisms responsible for adhesion has not prevented the phenomenal growth in commercially available adhesives. A two-volume catalog of adhesives [32] lists roughly 3000 different adhesives that are available from major United States manufacturers. To this list must be added adhesives from foreign manufacturers that are becoming increasingly available in the United States, as well as the products of small speciality companies. While there is a considerable overlap among many of the adhesives produced, there is also a great diversity in properties, characteristics, materials for which they are intended, curing conditions, temperatures of application and use, environmental conditions for which they are intended, and so on. Adhesives also come in a wide variety of forms and have widely differing application techniques or methods. When classified by mode of processing or cure, some common adhesive groupings are hot melt, anaerobic, cyanoacrylates (*instant* adhesives, sometimes called super glues), two-part curing, water based, solvent based, emulsion, contact, pressure sensitive, and film adhesives.

2 TESTING PHILOSOPHY

The failure of an adhesive joint is a complex phenomena that involves many factors including properties of the adhesive and the adherends, the nature of the interface (or interphase), the cure cycle, the rate of loading, temperature, humidity, and intricate details of geometry. Therefore, it is difficult to specifically define the strength or quality of a given adhesive. For example, despite the fact that various handbooks and manufacturers' literature often list the shear strength as if it were a well-defined property for an adhesive, it is important for the user to note that these values apply *only* for a given set of specific conditions. The *strength* values of the adhesive in joints that differ only slightly from joints tested under ideal testing conditions may differ markedly from the published values. Some of the reasons for these differences are clarified later in this chapter.

Testing is important in all aspects of mechanics and material design, and this is especially true for adhesive joints. It is generally recognized that such things as holes, threads, and notches associated with mechanical connectors such as bolts, screws, and rivets act as stress risers. Using a small round hole in a very large plate as an example, the best possible case is that the local stress at the edge of the hole will be increased by a factor of 3 over the nominal value in a similarly loaded plate with no hole. It is from such stress concentrations that crack growth often proceeds. The experienced engineer, when inspecting for fatigue cracks in a structure, devotes particular attention to the regions adjacent to such stress risers.

Adhesively joining parts can circumvent some of the problems associated with mechanical connectors by distributing the load over a larger area. This does not imply, however, that the stresses are uniform or that the stress distributions are well understood in adhesive joints. In fact, the regions near bond termini often act very much like the stress concentrations described earlier. Care must be taken to design adhesive joints correctly, just as it must be exercised when designing mechanical joints.

Adhesive tests are run for several reasons: (1) to compare two or more adhesives that are being considered for a given application; (2) to make certain that a given stock of adhesive has the same quality as previous shipments; (3) to ascertain whether the properties of an adhesive are currently the same as when originally received; (4) to compare the effects of different surface treatments, coupling agents, anodizing, and so on; and (5) to obtain parameters or *properties* that can be used to predict the strength of, or as an aid in the design of, practical joints to be used in a structure.

The first four reasons fall into a category of comparison or quality control tests, and it may appear, at least superficially, that almost any test can be used as long as care is taken to assure the test conditions are the same for all tests. There is, however, at least one fallacy to this line of reasoning. An adhesive that tests superior to another adhesive in one test may very well appear inferior if a different test is used. For example, some adhesives have relatively high butt tensile strengths but poor peel strengths, and vice versa.

It is generally more difficult to use results from standard laboratory tests to predict the strength of practical joints. As discussed in more detail later, the results of most adhesive tests are reported in the form of the force at failure divided by the bonded area. The maximum stress in the joint generally differs markedly from this average value. Since it can be argued logically that the initiation of failure in the joint is apt to be more closely related to the maximum stresses than to the average stresses, the average stress values at failure are likely to be of very limited use in design. It should be noted that the exact value of this maximum is often difficult to determine and is a function of the intricate details of the joint geometry, physics, and chemistry. Often the results from a test can *only* be used to predict the strength of other joints that *almost exactly duplicate* the test geometry in *every* detail. A goal of Section 4 is to illustrate how techniques currently being developed may ease this restriction.

Since moisture-induced failure across the interface may be the weakest feature of an adhesive bond, it is helpful to test the effect of moisture separately and select a primer that eliminates moisture-induced failure as a variable in bond testing. A simple accelerated test for screening silane primers for epoxy adhesives is described by Plueddemann [33]. In this test, a thin film of adhesive on a glass microscope slide or a metal coupon is cured and soaked in hot water until the film can be loosened with a razor blade. There is usually a clean transition between cohesive failure in the polymer and interfacial failure. Since the diffusion of water into the interface is very rapid in this test, the time–failure is dependent only on interfacial properties and may differ by several 1000-fold between unmodified epoxy bonds and primed epoxy bonds. Various silane primers differed by several 100-fold in performance with a given epoxy. In parallel tests, a thin film of epoxy adhesive on nonsilaned aluminum showed about the same degree of failure after 2 h in 70°C water as a silaned lap-shear joint in 150 days under the same conditions.

3   STANDARD TESTS

Several tests were formalized and standardized for evaluating adhesives. The American Society for Testing and Materials (ASTM) compiled the most complete description of such tests in [34], along with several other volumes containing tests for specific materials such as wood. These volumes can be found in most technical libraries; if not, one can contact the ASTM in Philadelphia, PA. Members of ASTM, and related agencies in other countries, perform a very valuable service in devising, designing, debugging, and publishing details for standard ways of testing and reporting results from tests, thereby facilitating comparisons between test results obtained in one laboratory with those obtained in another. The ASTM has also developed standards to evaluate many different aspects of adhesives. Included in the standards are ways to measure a variety of different properties of an adhesive. In this chapter we mainly concentrate on tests to determine mechanical properties, most notably, *adhesive strength*.

Adhesive strength tests can be characterized into three rather specific categories and one relatively recent approach that can be classified as a more general method. The three test categories are tensile, shear, and peel. After discussing these categories, a brief discussion of a few tests that are specifically designed to yield information that can be used in fracture mechanics analyses will be presented.

3.1   Tensile Tests

Given the choice, a designer seldom uses adhesives in a direct tensile loading mode. The primary reason for this is probably related to the fact that by overlapping, scarfing, and so on, the contact area can be increased. Figure 5.1 compares a butt joint with several other joint geometries that result in increased adhesive area and that also transform some of the tensile loading to shear. Another reason, based on experience, involves the empirical observation that a great many adhesives demonstrate poor strength and high sensitivity to alignment when exposed to butt-type tensile stresses. Nevertheless, several adhesive tensile tests are in common use. Figure 5.2 illustrates a few of these, with the most common being those described in [35]. Figure 5.2a shows two of the sample configurations from [35] that are suggested for measuring the strength of metal–metal and wood–wood bonds. Note that the sample's diameter for this geometry is chosen so that it yields a cross-sectional area of $645\,\text{mm}^2$ ($1\,\text{in}^2$). This test is often termed the *pi test*.

The specifications for U-shaped grips to help maintain alignment during loading for these samples are described in detail in [35]. It has been our experience, however, that even with such grips and reasonable care in manufacture, machining tolerances and other factors make it difficult to apply *axially centric* loads. As a consequence, data from such experiments often exhibit relatively large amounts of scatter unless extraordinary care is exercised in

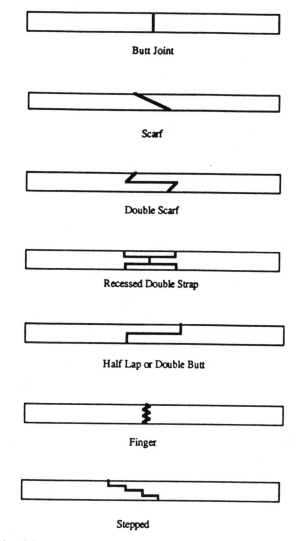

Butt Joint

Scarf

Double Scarf

Recessed Double Strap

Half Lap or Double Butt

Finger

Stepped

Figure 5.1 The butt joint compared with several other joints specifically designed to increase area and/or reduce the tensile stress component and increase the shear stress component in the adhesive bond.

manufacture, in alignment of the samples, and in performing the experiments [36].

Other standard tensile tests use samples that are easier to manufacture than those of Figure 5.2a. In [37] preparation of bar and rod specimens for butt tensile testing is described. Half-specimens of the recommended geometries are shown in Figure 5.2b. These samples are loaded by pins through 4.76-mm holes, and even though [37] describes a fixture to assist in sample alignment, eccentric

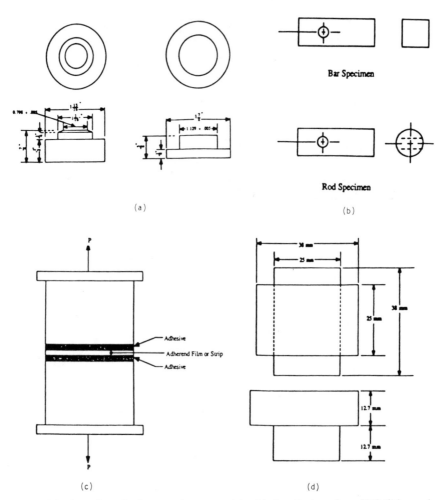

Figure 5.2 Several standard test specimen geometries: (a) pi peel test specimen [35], (b) bar and rod tensile specimens [37], (c) tensile test specimen for sandwich constructions [38], and (d) cross-lap specimen [39].

forces still generally pose problems in such tests even greater than those in [35]. The specimens of Figure 5.2a and b can be adapted for use with materials that cannot be readily manufactured into the conventional specimen shapes by the method shown schematically in Figure 5.2c. Examples of such materials are glass, ceramics, thin polymers and other thin films, and micalike materials. To test the *tensile strength* between an adhesive and one of these materials, the tensile strength must obviously be less than the tensile strength between the adhesive and the metal from which the end spools are manufactured. In [38] sample configurations similar to that shown in Figure 5.2c for investigation of the tensile strength of sandwich constructions in the flat plane are described.

Reference [39] describes a cross-lap specimen of the type shown in Figure 5.2d for determining tensile properties of adhesive bonds. Wood, glass, sandwich, and honeycomb materials have been tested as samples in this general configuration. Even under the best of circumstances, one would not anticipate that the stresses in such a case would be very uniform. The exact stress distribution is highly dependent on the relative flexibilities of both the cross beams and the adhesive. Certainly, caution must be exercised when comparing tensile strength from this test with data obtained from other tensile tests.

The results from these tests are normally reported as the force at failure divided by the cross-sectional area. Such average stress information can be misleading. The importance of alignment has already been discussed. Even when alignment is perfect and the bonds are of uniform thickness over the complete bond area, the maximum stresses in the bond line can differ markedly from the average stress. The distribution of stresses along the bond line is a strong function of the adhesive joint geometry. This is demonstrated in Figure 5.3, which shows stress distribution as a function of position for specimens of the general shape shown in Figure 5.2c. This family of curves was calculated by finite element techniques [40]. These calculations assume an elastic adhesive with a Poisson's ratio of 0.5, which is much less rigid than the steel adherends. The different curves in the family represent differing adhesive thickness/diameter

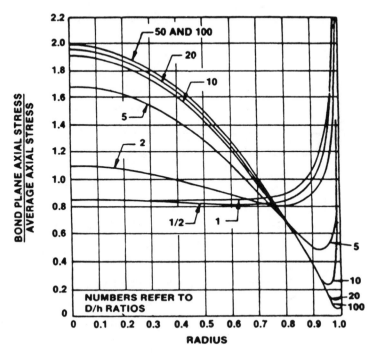

Figure 5.3 Stress distribution as a function of position at bond line for butt joint specimens for different thickness-to-diameter ratios (Poisson's ratio = 0.50).

ratios, shown as the parameters near the curves. Later, these results will be used to show how appropriate analysis of the stresses and strains in the adhesive can be used to predict strength and to predict the locus of initial crack growth.

In conclusion, the usual (average) results often reported from standard tensile tests *must be used with extreme caution* in predicting the strength of different (even if superficially similar) joints.

3.2   Peel Tests

Other tests in common usage are the peel tests. Figure 5.4 shows three types of peel specimens. We can understand ASTM D1876 [41] by examining Figure 5.4a. In this adhesive peel resistance test, often called the T-peel test, two thin 2024T3 aluminum sheets, 152 mm wide by 305 mm long in size, are bonded over a 152-mm-wide by 241-mm-wide area. The samples are then usually sheared or sawed into 25-mm-wide by 305-mm-long strips. At other times, the sample is tested as a single piece. The 76-mm-long unbonded regions are bent at right angles, as shown in Figure 5.4a to act as tabs for pulling with standard tensile testing grips in a tensile testing machine.

In a related test, one of the adhering sheets is either much stiffer than the other or is firmly attached to a *rigid* support. Various jigs have been constructed to hold the stiffer segment horizontal and, by using rollers or other means, to allow it to "float" so as to maintain the peel point at a relatively fixed location above and below the grips and at a specific peel angle, as shown schematically in Figure 5.4b (from [42]).

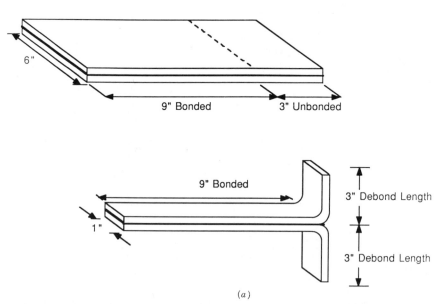

(a)

Figure 5.4   Some standard peel test geometries: (a) T-peel test specimen [41], (b) floating roller peel specimen [42], and (c) climbing drum peel specimen [43].

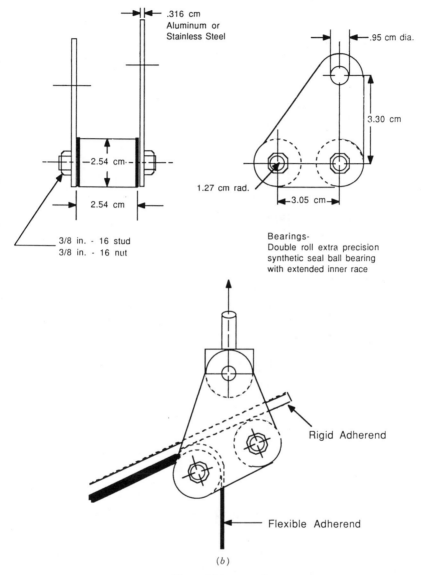

.316 cm
Aluminum or
Stainless Steel

2.54 cm

2.54 cm

3/8 in. - 16 stud
3/8 in. - 16 nut

.95 cm dia.

3.30 cm

1.27 cm rad.

3.05 cm

Bearings-
Double roll extra precision
synthetic seal ball bearing
with extended inner race

Rigid Adherend

Flexible Adherend

(b)

Figure 5.4 (*continued*)

Reference [43] describes the *climbing drum peel test* that incorporates light, hollow drums in spool form with each end having a diameter larger than the central portion, as illustrated in Figure 5.4c. Flexible straps are wrapped around the larger diameter spool and attached to the grips of a loading machine, and the flexible part of the peel specimen is wrapped around the smaller spool. Upon loading, the flexible straps unwind from the drum as the peel specimen is wound

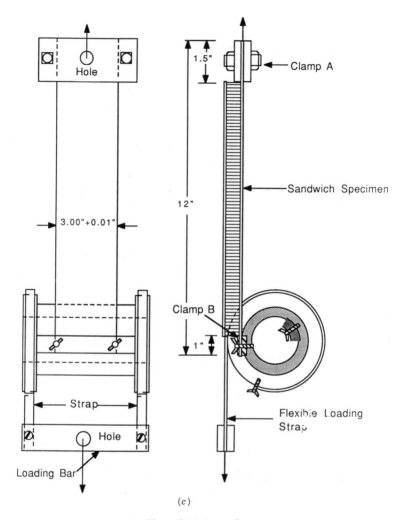

(c)

Figure 5.4 (*continued*)

around it and the drum travels up (hence the name climbing drum peel test) thereby peeling the adhesive from its substrate.

One of the simplest peel tests to conduct is the 180° peel test described in [44]. In this test, one adherend is much more flexible than the other so that upon gripping the two ends, during peel they assume the configuration shown in Figure 5.4d.

Clearly, peel strength is not an inherent fundamental property of an adhesive. The value of the force required to initiate or sustain peel is not only a function of the adhesive type, but it also depends on the particular test method, rate of loading, nature, thickness of the adherend(s), and other factors. Regardless, the peel test has proven to be a useful test.

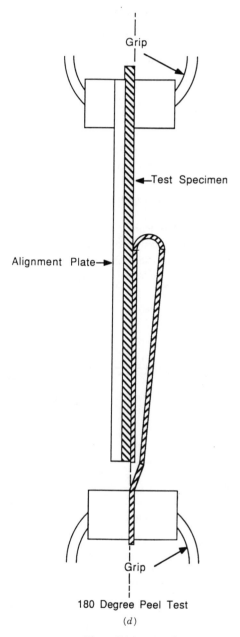

180 Degree Peel Test

(*d*)

Figure 5.4 (*continued*)

We are particularly impressed with very interesting studies conducted by the researchers at Akron University [45–47] who used peel tests to great advantage to measure the work of adhesion (see Section 4). Using the concepts of fracture mechanics in conjunction with the peel test, Gent and Hamed [45, 46] obtained insights into time–temperature effects, the role of plasticity, and many other aspects of adhesive fracture.

3.3 Shear Tests

Without doubt, the most commonly used adhesive test is the lap shear test. This occurs because samples are simple to construct and appear to resemble the geometry of many practical joints closely. Again, the stress distribution is not uniform, although it is conventional to report the results as the load at failure divided by the area of overlap. Here, too, the maximum stress can differ markedly from this average value. Furthermore, we will see that failure is more closely related to the value of the induced tensile stresses than it is to the shear stresses. Figure 5.5 shows some of the more commonly used lap shear specimen geometries. Figure 5.5a shows the geometry recommended in [48], which is the most commonly used.

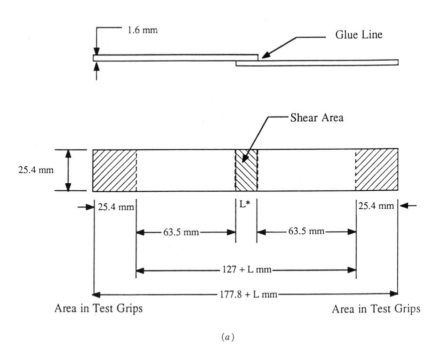

(a)

Figure 5.5 Standard lap shear specimens: (a) single lap shear specimen [48] or [49], (b) single lap shear specimen (*L = length of test area; length of test area can be varied; recommended length of lap is 12.7 + 0.3 mm) [49], and (c) double lap shear specimen (T1 = 1.6 mm, T2 = 3.2 mm, A = test gluelines, B = area in test grips, C = shear areas).

Since the flow of the adhesive out of the sides and the edges of the overlap often poses a problem, the samples are sometimes prepared from two relatively large sheets and the specimens are cut from the resulting laminated sheet. This is illustrated in Figure 5.5b and described in [49]. It has long been recognized that

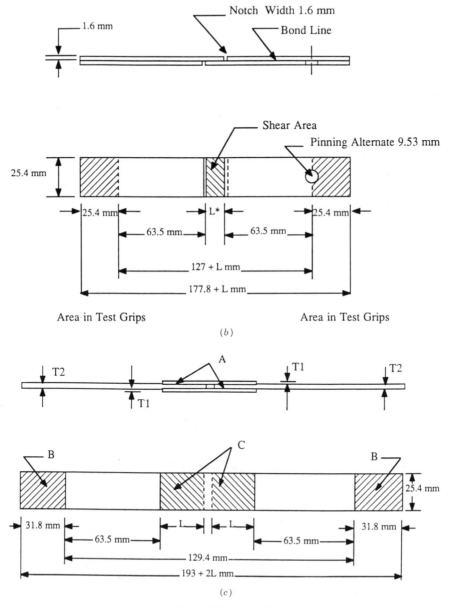

Figure 5.5 (*continued*)

the samples represented by Figure 5.5a must distort so that the forces applied to either end of the sample fall on the same line of action. This distortion induces cleavage (tensile) stresses near the bond termini. Efforts to alleviate this problem led researchers to develop a double lap shear specimen as shown in Figure 5.5c (from [50]).

That these efforts have not been completely successful should be clear from the photograph in Figure 5.6, which shows a copper adherend-epoxy adhesive double lap shear specimen after loading to failure. The extensive outward bending of the cover plates provides graphic evidence that the adhesive and supporting plates were not subjected to shear stresses only.

Reference [47] describes efforts to analyze the stresses in a lap shear specimen with the aid of elastic finite element techniques and to explain some of the phenomena associated with lap shear joint failure. These analyses demonstrate that the stresses (both shear and tensile) are very nonuniform because of adherend bending and the mathematical singularities (in the elastic case) near the bond termini. The results of these analyses are supported by experimental studies. These results, as well as similar results by others [51], indicate that failure of lap shear specimens is dependent more on the value of the cleavage stresses near the bond termini than on the maximum shear stress, let alone on the average shear stress generally reported in handbooks, manufacturer's literature, and so on.

While the categories of tests described previously are most familiar for testing the strength of adhesives, they are by no means all inclusive. There is a wide variety of other standard tests available to determine tensile, shear, and peel properties, as well as other characteristics. Some of these tests are extremely useful in screening potential candidates for specialized adhesive applications. The so-called Boeing wedge test [52] has proven useful for investigating the

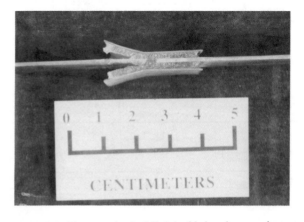

Figure 5.6 Photograph of a failed double lap shear specimen.

environmental (moisture–time) stress susceptibility to failure of adhesion. Figure 5.7 shows the experimental configuration used by McMillan and his associates [53].

References [54–56] describe equipment and techniques that can be used to investigate performance under sustained loading with sample configurations similar to those previously described. Such creep apparatus often incorporates springs to maintain a load on a sample over long periods of time.

Reference [57] describes apparatus that can be used to explore the failure of adhesive samples at high temperatures by radiant heating using tubular quartz lamps, while [58] outlines procedures for testing samples at low temperature (-268 to $-55°C$). The American Society for Testing and Materials also provides specific standards (or recommended practices) to investigate properties

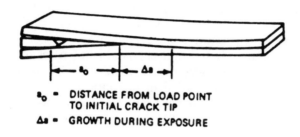

a_o = DISTANCE FROM LOAD POINT TO INITIAL CRACK TIP

Δa = GROWTH DURING EXPOSURE

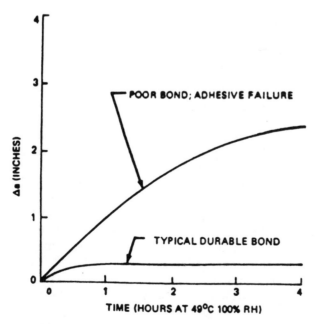

Figure 5.7 The Boeing wedge test and typical results.

of adhesives that are less directly related to mechanical strength. Such features include resistance to mold attack [59], chewing by rodents [60], eating by insects [59], and so on.

Other tests attempt to infer long-term behavior from short-term tests. While such accelerated tests are never perfect, they may be the only alternative to observing a part in actual service for decades. The latter alternative is, of course, unsatisfactory if it involves waiting for this period of time to put a product on the market. As a case in point, consider the boiling test [61]. It is recognized that an adhesive for laminating wood will, in all probability, not actually be exposed to boiling in water. It is hoped, however, that resistance to boiling for a few hours or days may provide some valid evidence (or at least insight) into the durability of a laminated wood panel after years of exposure to high ambient humidity and temperature. Another accelerated aging test is described in [62]. This test determines the deterioration in the strength of adhesives exposed to high temperatures and high-pressure oxygen (typically 70°C and 2.07 MPa), again in an effort to at least provide screening in order to identify likely candidates for more practical aging conditions.

Surface preparation is an extremely important part of adhesive science and technology. This often includes much more than just cleansing a surface. It may involve roughening and preparing an appropriate oxide or other surface layer for receiving the adhesive and perhaps facilitating mechanical interlocking. Techniques standardized by ASTM include [63] and [64] for metals and [65] for plastics.

4   ADHESIVE FRACTURE MECHANICS

It should now be clear that the stress distribution in adhesive joints is complex. Furthermore, the details of this distribution are highly dependent on specific details of the system. Such details include geometry, relative moduli of adhesive and adherends, adhesive and adherend thicknesses, and so on. In addition, the maximum stresses in the bond almost always differ markedly from the average value. Indeed, elastic analyses often exhibit mathematical singularities at geometric or material discontinuities. From these observations, it should be clear that the use of the conventionally reported results from most tests (i.e., values of the average stress at failure) would be of little use in designing joints that differ in any significant detail from the sample test configuration.

In such cases, the concepts of fracture mechanics have much to offer. One of the more popular and graphically appealing approaches to fracture mechanics views the joint as a system in which failure (often considered as the growth of a crack) of a material (or joint) requires: (1) the stresses at the crack tip to be sufficient to break bonds and (2) an energy balance. In this model, it is hypothesized that even if the stresses are very large (often theoretically infinite in the elastic case), a crack can grow *only* if sufficient energy is released from the stress field to account for the energy required to create the new crack (or adhesive debond) surface as the fractured region enlarges. The specific value of this energy (in joules per square meter of crack area) is given various names in the literature; for example, fracture energy, adhesive fracture toughness, and work of adhesion. The word adhesion is dropped from the comparable term when cohesive failure is being considered. This embodiment of fracture mechanics involves both a stress–strain analysis and an energy balance.

The analytical methods of fracture mechanics (both cohesive and adhesive) are described in [66, 67]. These are not repeated here; however, a brief outline of a numerical approach that can be applied is described to provide the reader with some insight into the concepts, principles, and methodologies involved.

Inherent in fracture mechanics is the concept that natural cracks or other stress risers exist in materials and that final failure of an object generally initiates at such points [66, 67]. Let us assume, therefore, that a crack (or region of debond) is situated in an adhesive layer. Modern computation techniques are available (most notably, finite element methods) that facilitate the computation of stresses and strains throughout a body, even if analytical solutions may not be possible. The stresses and strains are calculated throughout the entire adhesive system (adhesive and all adherends), including the effects of a crack in the bond. These can then be used to calculate the strain energy, U_1, stored in the body for the particular crack size, A_1. Next, the hypothetical crack is allowed to grow to a slightly larger area, A_2, and the preceding process is repeated to determine the strain energy, U_2. This approach to fracture mechanics assumes that at critical crack growth conditions, the energy loss from the stress–strain field goes into the formation of the new fracture energy. The quantity $\Delta U/\Delta A$ is called the energy release rate. The so-called critical energy release rate $(\Delta U/\Delta A)_{crit}$ is that value of the energy release rate that will cause the crack to grow. Loads that result in energy release rates lower than this critical value will not cause failure to proceed from the given crack, while loads that produce energy release rates greater than this value will cause it to accelerate. This critical energy release rate value is equivalent to the adhesive fracture energy, or work of adhesion, previously noted.

This simple model of fracture mechanics should help the reader, unfamiliar with fracture mechanics, to visualize the concept of fracture, albeit this example is somewhat of an oversimplification. The molecular mechanisms responsible for the energy required to create a unit of new area, referred to as the fracture energy or fracture toughness, are not completely understood. It generally involves more than simply the energy required to rupture a plane of molecular bonds or, taking roughness into consideration, the energy required to form this increased area. In fact, for most practical adhesives, this energy is a small but essential fraction of the total energy. The total energy includes energy that is lost because of viscous, plastic, and other dissipation mechanisms at the tip of the crack. Hence, linear elastic stress analyses are inexact.

Fracture mechanics has found extensive use in design to minimize the probability of cohesive failure, but its use for analyzing failure of adhesive systems is more recent. There has, however, also been a significant amount of research and development in the adhesive fracture mechanics area. To review it all, even superficially, would take more space than is allocated for a chapter of the type considered here. We would, however, like to demonstrate how fracture mechanics provides insight into behavior and can be used to analyze problems that cannot be readily treated by *average stress* criteria. This will be accomplished by relying largely on the authors' and associates' own analysis and experimental work. There is no intention to slight the work of others or to intimate that other work is not as significant (or more so) as our own. Before embarking on a description of some of our own results, we will provide a short listing of researchers in the area who have done closely related work:

E. H. Andrews [68, 69], D. W. Aubrey [70], W. D. Bascom [71, 72], H. F. Brinson [73], D. Broek [74], J. D. Burton [75], G. Danneberg [76], F. Erdogan [77], A. N. Gent [78, 79], T. R. Guess [80], G. R. Hamed [81], R. W. Hertzberg [82], G. R. Irwin [83], W. S. Johnson [84], D. H. Kaelble [85], H. H. Kaush [86], A. J. Kinloch [4, 87], W. G. Knauss [88], K. M. Liechti [89], J. C. McMillian [90], S. Mostovoy [91], D. R. Mulville [92], J. R. Rice [93], E. J. Ripling [94], E. F. Rybicki [95], G. B. Sinclair [96], J. D. Venables [97], S. S. Wang [98, 99], J. G. Williams [100], M. L. Williams [101], and R. J. Young [102].

Not only is this listing incomplete, but many of the researchers have scores of other publications. We have only listed one or two for each. Hopefully, this will provide the reader with a starting point from which more details can be found from reference cross listings, searching of citation indexes, abstracting services, and so on.

Among other topics, the preceding researchers have treated such subjects as: theory [4, 68, 70, 77, 85, 87, 92, 98–101, 103]; mode shape, thickness, and other geometric dependence [71, 86, 93, 94, 98]; plasticity and other nonlinearities [4, 71, 79, 86, 98, 99]; numerical methods [71, 90]; sample geometry [75, 76, 81, 86, 89, 90, 98, 101]; testing techniques, and so on [69, 84, 92, 95, 97]; ruthem adhesives [70, 79]; rate and temperature effects [78, 79, 86]; fatigue [82, 91, 102]; failure of composites [104]; as well as a wide variety of other factors and considerations in adhesion.

Modern finite element or other numerical methods have no problem in treating nonlinear behavior [74, 100–102]. Our physical understanding of material behavior at such levels is lacking and effective use of the capabilities of such computer codes depends, to a large extent, on the experimental determination of these properties. For many problems, it has become conventional to lump all dissipative effects together into the so-called fracture energy and not be overly concerned with separating this quantity into its individual energy absorbing components.

Several researchers (some of whom were just cited) have enjoyed significant successes in explaining many aspects of adhesive performance and predicting the strength of a bond from tests on other, quite different, joints by using linear elastic fracture mechanics for adhesives for which the stress–strain behavior is fairly linear. Another fracture mechanics approach, called the J-integral, has some advantages in treating nonlinear behavior [74, 93, 100].

Researchers in numerical analysis have developed many labor and computer timesaving techniques to aid in the solution of problems of the type outlined earlier. For example, calculating the energy release rate as described earlier is quite helpful in describing and visualizing the approach. However, its use has become uncommon in actual calculations because such approaches as the crack closure method (CCM) are much more "efficient" [105]. In this method, a finite element grid is established throughout the specimen, usually with finer elements near regions of high stress (e.g., near crack tips). The nodes of these elements extend along the free surfaces of the crack. The energy release rate is inferred by calculating the energy associated with the forces (shear and normal) and the displacements required to bring the first node from the crack tip back into conjunction with its matching node of the complementary surface; that is, the crack is closed over one finite element of its length. The absolute value of the energy required for this operation divided by the area associated with the closure is taken as the energy release rate. Analysis showed that when appropriate care is taken (grid size and other considerations), the numerical results obtained by the crack closure method agree very well with those obtained by the more direct approach [40, 106]. This method also facilitates partitioning of the energy into tensile (mode I) and shear (mode II) stresses at the crack tip.

It was previously noted that most adhesive systems are not linearly elastic up to the failure point. Nevertheless, researchers showed that elastic analyses of many systems can be very informative and useful. Several adhesive systems are sufficiently linear so that it is possible to lump the plastic deformation and other energy dissipative mechanisms at the crack tip into the adhesive fracture energy (critical energy release rate) term.

We would like to describe briefly some of our own finite element analysis results that we feel are informative. In the first of these [40], a tensile butt joint specimen with a soft rubbery adhesive {modulus of elasticity, $E = 3.5\,MPa$ (500 psi) and Poisson's ratio $v = 0.5$) is used to bond a relatively rigid plastic [$E = 350\,MPa$ (500,000 psi)] circular rod} to a similarly rigid plate substrate.

A finite element analysis of this system was undertaken with an assumed small crack at either the edge or at the center of the circular bond region for various adhesive thickness-to-diameter ratios t/d. When energy balances of the type described earlier were performed, it was determined that for some adhesive thickness-to-diameter ratios, the energy-release rate was greater for flaws located at the sample center than for the same size flaws located at the edge. For other ratios, the opposite was true. Furthermore, the analysis allowed us to predict the effect of the t/d ratio on the relative strength of the joints. Stated another way, if the strength at one t/d ratio is known (or assumed), the analysis allows us to calculate the energy release rate associated with that load for a given flaw size. Then, assuming that the same manufacturing techniques and other factors result in comparable flaws at other thickness-to-diameter ratios, the strength at other thicknesses can be calculated and the locus of initial failure can be predicted.

To check this hypothesis, the following experimental setup was designed and constructed. Using a model polyurethane adhesive (Thiokol–Solithane 113), 38-mm- (1.5-in.-) diameter rods of clear PMMA were bonded to PMMA sheets. These samples were mounted in an Instron tensile testing load frame in such a way that a light shone at the top of the rod would allow a video camera to observe the bottom of the adhesive clearly through a mirror arrangement. Care was taken to load the specimen in as near pure axial tension as possible, determined by mounting strain gauges (on selected samples) at 120° intervals around the rod part of the specimen. The results from these tests are shown in Figure 5.8. The finite element analyses indicated that for a t/d less than about 1.2 (for samples of these materials and geometry), the energy release rate for cracks at the center was larger than for those at the edge. For a $t/d > 1.2$, the opposite was true. In Figure 5.9 the open data points are those for which the video pictures indicated crack growth at or near the sample center, while the solid data

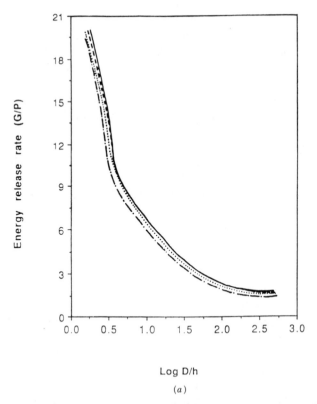

(a)

Figure 5.8 Energy release rates for edge and center flaws in a butt joint as a function of the diameter-to-thickness ratio for various flaw sizes and assuming $v = \frac{1}{2}$: (a) flaw at center (flaw size: ————, 0.0037 D; ----, 0.0027 D; ...., 0.0017 D; ·—·—·—, 0.0012 D) and (b) flaw at edge (flaw size: ——, 0.0055 D; ----, 0.0035 D; ...., 0.0025 D; ·—·—·—, 0.0015 D).

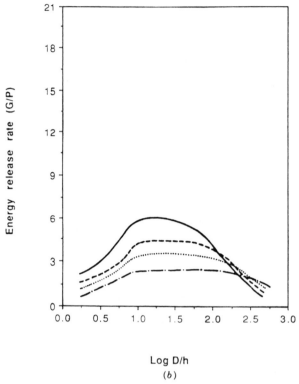

Figure 5.8 (*continued*)

points are for cracks that grew from the sample edge. It can be seen that these data agree quite well with the finite element predictions. Furthermore, the data points are nicely grouped along the two curves, again in good agreement with the finite element predictions. In the complete study, a variety of different inherent flaw sizes were analyzed. While this affected absolute values, the trends were in every case similar to those shown in Figure 5.9.

In a second series of analyses and associated experiments, the inherent flaw sizes of a given adhesive [40] and sample preparation techniques were determined by first introducing crack regions of known sizes and then extrapolating the data back to the load for nonpurposely introduced cracks. It was assumed that failure in the nonartifically cracked specimens initiated as inherent microcracks, the size of which could be determined by extrapolation and by using the adhesive fracture energy determined from the previous tests in which large, deliberately introduced flaws were present. It was found that the value of this inherent flaw, along with the critical energy release rate for the adhesive, could then be used to predict strength of joints prepared from the same adhesive with the same surface preparation techniques but exposed to different types of loadings.

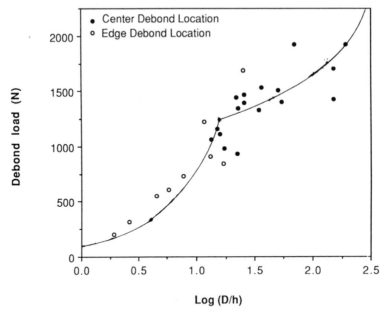

Figure 5.9   Effect of diameter-to-adhesive thickness ratio on location and load for bond failure in butt joints with a rubbery adhesive and relatively rigid substrates.

In one series of analyses and experiments, the energy release rate inherent flaw size relationship was determined for a given epoxy-steel adherend and surface preparation by testing of butt tensile specimens. Next, a series of analyses and experiments was conducted on blister specimens. In these experiments, extraordinary efforts were taken to keep as many factors as possible the same (e.g., the same metal was used for the adherend, its surface was prepared in the same way, and the same adhesive was used). The experiments were performed by pressurizing with air a circular region of debond (a blister) to failure. An energy balance of the type described earlier allowed the prediction of the required failure pressure. A series of experiments for various blister sizes agreed with numerical predictions, based entirely on parameters determined from the butt tests, to within 6% [107, 108].

In our opinion, an even more convincing argument for fracture mechanics was the finite element predictions of the failure loads for simple lap shear specimens constructed of the same type of steel with the same adhesive and sample preparation techniques as for the samples just described. We noted earlier that while this is commonly termed a shear specimen, near the bond termini the tensile stresses (and associated energy release rates) really dominate the mechanical response of the samples. Again, using a finite element calculation of the energy release rate and using the same critical energy release rate and inherent flaw size relationships previously determined in the tensile butt specimen, it is possible to predict the strength of the joint. Here, too, subsequently conducted experiments yielded failure loads to within 5% of the predicted values [107, 108].

5 CONCLUSIONS

The adhesive researcher or technologist has many standard test methods from which to choose. These are designed with various goals and objectives in mind. Many of these are useful for the purpose of comparing adhesives under different loading, investigating different types of chemical or physical attacks on adhesives, exploring aging phenomena, determining the effects of radiation and moisture combined with sustained loading on adhesive properties, and so on. On the other hand, care should always be exercised not to use the test results for purposes for which they are not well suited. Results from many of the adhesive strength tests are conventionally reported as the failure force divided by the bond area. Such average stress at failure results cannot, in general, be consistently and reliably used to predict failure of other joints that differ even slightly from the test geometry. Fracture mechanics approaches, on the other hand, show promise (and have, in fact, been used) to predict the strength of joints that differ considerably from the reference joint. Reference [109] describes a standard adhesive fracture mechanics joint in the form of a tapered double cantilever beam. The specimen dimensions are shown in Figure 5.10. We hasten to note, however, that fracture mechanics is not limited to this or any other

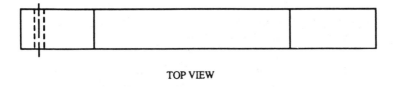

TOP VIEW

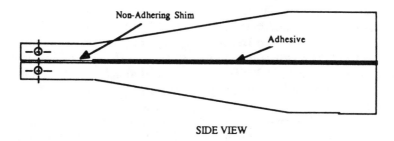

SIDE VIEW

Figure 5.10 Tapered double cantilever fracture mechanics specimen [109].

specific testing geometry. In principle, any geometry for which the above-described energy balance (or alternatively, calculation of the stress intensity factor, J-integral, etc.) can be accomplished can be used as an adhesive test.

Sometimes, circumstances dictate the use of a nonstandard test geometry. For example, a few years ago we were given the problem of measuring the quality of natural barnacle adhesive. The barnacle dictated the exact form of the joint between the base of the steel and the plastic sheets we had placed in the ocean. This form did not lend itself to tensile, lap shear, or split cantilever testing. It was, however, possible to predrill holes in the plate, and to fill these with dental waxes that were solid and relatively hard at the ocean temperatures near San Francisco, CA, where the barnacle growth experiments were conducted. The wax was later easily removed at moderately elevated temperature. The base of the barnacle covering this hole was thereby exposed and could be tested by application of fluid pressure, thus forming a blister. Measurement of the pressurization at failure allowed the determination of the adhesive fracture energy [9].

Once the adhesive fracture energy is determined by testing, fracture mechanics points the way that it, along with a knowledge of the flaw size and a stress–strain analysis of the joint, can be used to predict the performance of other joints. Modern computational techniques greatly facilitate the application of these methods.

Finally, it should be noted that the stresses, strains, fracture energy, and other such parameters used in the analysis may depend on loading rate (or time of loading), the mode of stress at the crack tip, temperature, and environment. Means of incorporating these parameters have been or are being developed.

6   APPENDIX: TYPICAL EXAMPLES OF PROCEDURES FOLLOWED FOR STANDARD TESTING

Mechanical tests require the use of a mechanical testing machine to apply tensile, compressive, bending, or shear forces to the sample. These typically fall into one of three classes. The first and simplest types are the *universal testing machines* comprising a load frame to which is affixed a hydraulic cylinder. The photograph in Figure 5.11 shows such a universal testing machine with a capacity of 27,215 kg (60,000 lb). Although this machine was manufactured by the Tinius Olsen Company in the mid-1950s, it is still very functional. In such equipment, the cylinder is driven by hydraulic pumps (usually electric, but sometimes hand operated); and it in turn causes a crosshead to move relative to another fixed (but usually positionable) crosshead attached to the frame. The sample is affixed to the crossheads by grips, or in the case of compression, bending, and so on, between platens; and it is loaded by moving the crossheads. The value of the applied load is determined by a load cell, which is sometimes no more complex than a pressure gauge in the hydraulic circuit. More often, an electronic or pneumatic load-sensing link is in the load string. Such systems have

Figure 5.11   An older model universal testing machine 27,215-kg (60-lb) capacity, manufactured by Tinius Olsen Company.

the advantage of being simple, relatively inexpensive, and available with very large load capacities. Major disadvantages are that they provide little or no control of loading rate and they usually have fairly slow response times. A few inexpensive machines of this type, with low to moderate loading capacity, use compressed air or bottled gases to drive a pneumatic cylinder affixed to a simple load frame. Other models of universal testing machines replace the hydraulic cylinder loading systems with screws that are driven at a constant rate.

The screw drive machine has perhaps been popularized most by the Instron Company. Because of the popularity of this company's equipment, such testing is sometimes generically classified as Instron testing, regardless of the manufacturer (of which there are several). The Instron Company also manufactures a variety of other testing equipment including servosystems, which are described later. This equipment again includes a loading frame with crossheads. The photograph in Figure 5.12 shows a typical 1985 vintage machine of this type with a capacity of 453.6 kg (1000 lb). Similar, but updated, versions by Instron (and other manufacturers) of varying capacities are very popular laboratory equipment today. Many are now equipped with computer control and computer data acquisition and analysis systems. Here, the moving crosshead is driven by twin screws. Electrical controls facilitate the turning of these screws at different "fixed" rates so that the samples can be tested at different preselected deformation rates. One of the crossheads has affixed to it an electronic load cell that drives one axis of a recorder or provides the input to a data acquisition system. The other recorder axis is commonly driven at a fixed time rate but may also be driven by other electrical transducers such as strain gauges and clip-on displacement gauges.

By their basic nature, these machines facilitate testing at constant displacement rates, but manufacturers have incorporated a wide variety of useful

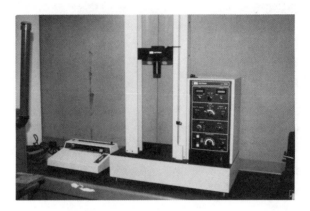

Figure 5.12   A 453.6-kg (1000-lb) Instron Incorporated testing machine.

features and options. Through these, the machines can be made to cycle between extremes in load or displacement, apply a fixed displacement to determine the load as a function of time (i.e., a relaxation test), and so on. A wide variety of grips are available to apply loads to various materials (e.g., flat or round metals, elastomers, fibers, and composites). Likewise, there are standard jigs for conducting bending tests, shear tests, and other specialized tests.

Since many standard adhesive strength tests specify that the loading be done at a fixed displacement rate, Instron-type testing machines are well suited for the adhesive testing laboratory. They are typically moderate in cost, simple to operate, and very durable. They do not, however, have the versatility of servocontrolled equipment.

Servocontrolled testing equipment, an example of which (manufactured by MTS Incorporated) is shown in the photograph of Figure 5.13, can be programmed to apply a wide variety of loading histories to a sample, from very simple to very complex. The system shown in this figure is again a load frame with a fixed crosshead. The electronic load cell is usually attached to this crosshead. One grip, compression platen, or other fixture is then connected to this load cell. The complementary grip is attached to the servocontrolled loading element. This element is usually a piston attached to a hydraulic cylinder, but it can also be an electronically controlled screw or other mechanical device. The motion of the hydraulic cylinder is controlled by a servoamplifier controller through servovalves. The servocontroller is basically a computer (in the past these have been primarily analog, but now the controller is often digital) that sends signals to a servovalve, which causes it to induce a prescribed motion in the hydraulic cylinder or, in the case of a mechanical system (i.e., screw), a prescribed operation of a servomotor. As an example, if it is desired to subject a sample to an alternating load, an electrical signal from the load cell as well as an alternating control signal would be fed into the

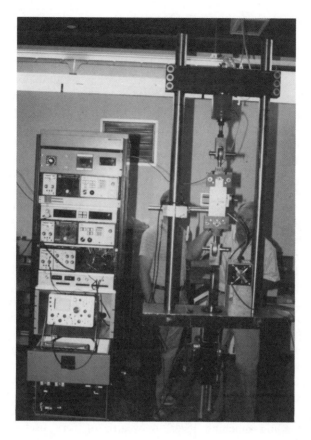

Figure 5.13 A MTS Incorporated servocontrolled testing machine.

servocontroller. The electronics of the control system forces the load cell signal to follow the control signal. This is done in the following way: the electrical signal from the load cell is proportional to the value of the load applied to the sample. The controller compares this with the desired value of the control signal. If these differ, it sends a signal to the servovalve that in turn directs high-pressure oil to the side of the cylinder that causes it to move to bring this error signal to zero. In a properly designed system, this response is so rapid that the load on the sample almost exactly duplicates the command dictated by the control signal. If it is desired to control the strain in a sample or the displacement of the sample grip, the signal from the load cell can be replaced by the output from a resistance strain gauge bonded to the sample or a displacement transducer connected between the grip supports. In this way, the sample loading history that can be generated (or simulated) is limited electronically only by the dynamic responses of the system (e.g., the hydraulic power supply must be able to provide enough oil at sufficient pressure and flow volume to move the system rapidly enough to respond to the given command). One can, for example, conduct a constant displacement rate test (Instron test) on a sample by utilizing a displacement transducer attached to the hydraulic cylinder shaft and using the electrical output of a ramp generator as the control signal.

The obvious advantage of such a system is that it cannot only be used to conduct rather simple tests such as the constant displacement test just mentioned, but it can also be used for more complex loading such as that required for investigation of material fatigue. The control signal can be sinusoidal, sawtooth, square wave, or of a more complex form. Within the dynamic limits of the machine, the sample can be subjected to a sinusoidal, sawtooth, and so on force (or displacement or strain, depending on the type of response transducer employed). The servocontrolled testing system is, therefore, an extremely versatile instrument. Its major disadvantage lies in its cost, which starts at roughly twice that of the simpler (constant displacement) machines of the same capacity. Its use is also somewhat more complicated, requiring more skill and care on the part of the operator, because the system responds very quickly and can present some safety problems in the hands of unskilled or careless operators. Hydraulic power supplies are also somewhat noisy in a rather annoying part of the sound spectrum. This problem can often be alleviated, but not totally eliminated, either by good sound insulation or by locating the power supply remotely.

Environmental chambers are available for use with all of the mechanical testing equipment described that facilitates testing at temperatures from near liquid nitrogen temperatures of about $-180°C$ up to $500°C$ or more, as well as subjecting the loaded sample to various gaseous environments. In all cases, using appropriate grips and loading fixtures is as important as the mechanical loading machine itself. A wealth of very useful general and specialized accessories are available from testing machine companies or specialized manufacturers. Most well-equipped mechanical testing laboratories are likely to have as much invested in accessories as in the basic testing machines.

To those who have never performed an adhesive strength test, the procedures may look deceptively simple. It is emphasized that to obtain results it is essential that samples be prepared and tests conducted with great care and diligence, according to prescribed specifications. To help develop an appreciation for the difficulties and pitfalls involved in this type of testing, the procedures involved in three different types of tests will be described: the button tensile test, the lap shear, and peel tests.

In the first of these, one must first manufacture the buttons and the loading fixtures. The buttons are manufactured from a material whose tensile joint strength with a given adhesive is desired. For a series of tests in this laboratory, the buttons were made of a high-strength steel (the same as used in some parts of the space shuttle) heat treated to a tensile strength of 220 ksi. For the tensile test the surface must not only be bonded to the machined surface and receive the appropriate surface finish and treatment, but a means of applying the tensile load to the specimen must also be devised. In our studies, a circumferential groove is machined as shown in the schematic drawing of Figure 5.14 (also, see [35]). For reasons to be discussed later, it is absolutely essential that this groove be very precisely machined and that the bonded surfaces be plane and parallel. Almost all standard test procedures (e.g., ASTM) require that several test samples (usually at least five) be tested for each test condition. A number of such buttons would therefore need to be manufactured. In our laboratories it is common to manufacture lots of 20 or more. These buttons require loading yokes through which the load is applied. The ones manufactured for use in our laboratories are shown in the drawing of Figure 5.14 and in the photograph of Figure 5.15.

For butt tensile specimens, the surface to be bonded must first undergo a prescribed surface treatment. It is absolutely essential that this surface be the same as that to be used in service. The surface treatment is generally as important to final bond strength as is the selection of the adhesive type. Surface preparation very often involves more than simple cleaning, and it is generally important to have the appropriate roughness. Commonly, chemical treatments are prescribed that develop an appropriate oxide layer to receive the adhesive.

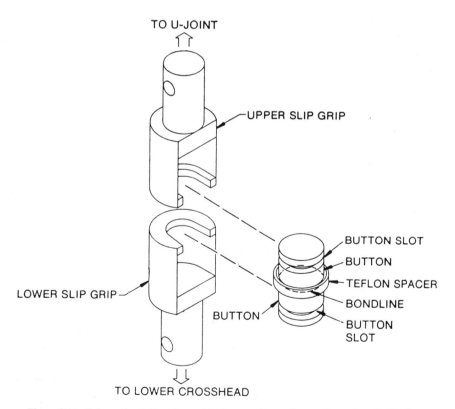

Figure 5.14 Schematic of slip grips and button specimens for tensile testing of adhesives.

Figure 5.15 Photograph of button tensile being inserted into the upper slip grip mounted in an Instron machine.

This may involve removing loose scale, dirt, and oxides and replacing them with a more tightly adhering layer. At times it is required that the appropriate layer is not only coherently attached, but also in possession of an open structure or porosity that can be penetrated by the adhesive to increase adhesion by mechanical means, commonly referred to as *hooking* or *mechanical interlocking*. In addition, the time between surface preparation and adhesive application is important, as are the storage conditions during this period. The cleanliness of the room and the dryness and cleanliness of the air where the samples are prepared can also be important factors.

After the surface treatment is completed, the adhesive joints are prepared. In each step, it is important to follow prescribed procedures. The way the adhesive is mixed (e.g., thoroughly, entrapping air, etc.) can have a profound effect on strength. The time between mixing and application, or the *pot life*, can influence joint strength. Cure conditions such as temperature and the pressure applied are also important factors.

Adhesive thickness, as well as uniform thickness and alignment, are extremely important, as discussed earlier, and are reinforced later. *These are not easy factors to control precisely.* At times we used Teflon or polyethylene strips or films to help control thickness. The adhesive we investigated exhibited nearly no adhesion to these plastics. The hope is, therefore, that they will not affect joint strength other than that caused by factors relating to alignment and thickness variations. The adhesive is then allowed to cure. Temperature and time of cure can affect the adhesive properties, and they can also affect residual stresses that in turn influence joint strength.

The assembled bonded buttons are next loaded into the testing machine. In the testing operation, machine alignment and yoke dimensions and alignment are of the utmost importance. After the samples are in a position to be pulled to failure, the load is applied at the prescribed rate, and the load-deformation behavior is recorded. Since most adhesives are polymers and most polymers are viscoelastic–plastic, the rate of loading is critical. Typically, a rate of 0.5 mm/min (0.02 in./min) is used for our tests, but several speeds can be used to investigate rate effects. Testing temperature and humidity are also important parameters. Unless otherwise required we maintain our testing laboratory at 22°C (± 1) and 50% (± 5) relative humidity.

A simple analysis will serve to illustrate the ramifications of various alignment considerations noted in the previous two paragraphs. In this simple example we consider a 0.25-mm-thick adhesive with a modulus that is low compared to that of the cylindrical adherends (e.g., epoxy with steel). Furthermore, assuming that because of the variation of (1) adhesive thickness, (2) mismachining of the sample faces, (3) the machining of slots in the cylindrical sample pieces, (4) the machining of the test grooves, or (5) some combination of these factors, one edge of the lip on the test grips meets the edge of the slot in tensile specimen, while the opposite lip misses its corresponding slot edge by 0.025 mm. This is not an unreasonable assumption, since this is about the typical tolerance maintained by a good machinist. When the load is applied, the matching segments of the sample and yoke meet at one edge and are loaded first. In fact, in the extreme case, it is possible that one side of the joint can experience a strain of several percent before the other edge senses any tensile stress. Since a typical epoxy yields at strains of a percentage or two, stress is indeed distributed very nonuniformly across the joint. A more careful analysis, but one that still neglects the stress singularities at the edges, suggests that loading an elastic adhesive, such as that just described, from the edge rather than from centric loading results in roughly four times as great a stress as in the former case. As a consequence, if care is not taken to reduce alignment and related problems, the data will exhibit considerable scatter. Standard deviations approaching 50% of the average failure load are commonly reported for relatively brittle adhesives. On the other hand, we have been able to reduce this to approximately 5% where extraordinary care is taken to minimize these effects.

The lap shear test is one of the most commonly used methods to evaluate adhesive strength. It may be informative to outline typical steps that can be followed in this procedure [48] complemented by [49]. To reduce the amount of adhesive flow out from their edges, the samples are prepared first in fairly large sheets and then the final samples are cut from this sheet as described in [49] and shown schematically in Figure 5.16. Two sheets of the appropriate thickness (6.5 mm thick in the photographs shown in Figure 5.17) were cut to form squares 47.5 × 47.5 mm (18 × 18 in.). These were then subjected to the prescribed surface treatment. In our study, this involved a sandblast treatment, followed by an air blast to remove grit and a cleaning with an appropriate solvent. The square sheets were then bonded together with the adhesive and put through the prescribed cure cycle. Again, we emphasize the importance of adhesive thickness, cure cycle, and other factors that need to duplicate exactly the conditions that will be present in the intended end use. We found that the use of narrow

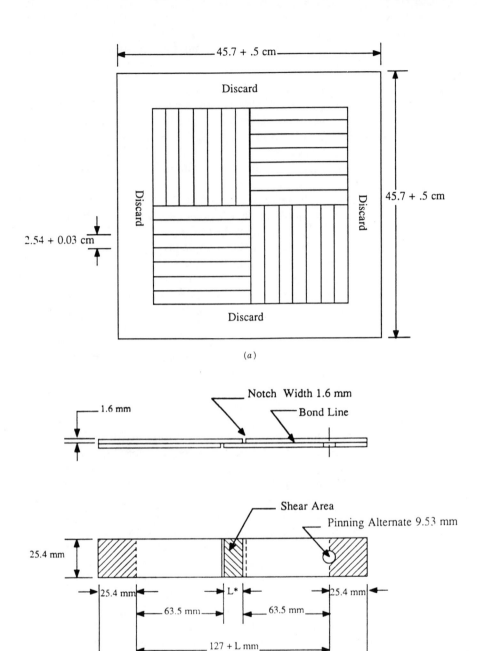

(a)

(b)

Figure 5.16 (a) The layout for cutting 28 lap shear specimens from two large sheets bonded together (all discards to be of common width as determined by width of cutting tool), and (b) the final dimensions and saw cut notches to produce a lap shear specimen (*L = length of test area; length of test area can be varied; recommended length of lap is 12.7 + 0.3 mm).

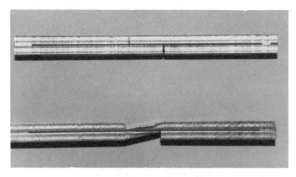

Figure 5.17 Photograph of lap shear specimen. The upper specimen is a standard lap shear specimen per [49]. The lower specimen has the metal over the overlap tapered to investigate effects of reducing the stress concentration factor at the saw cut.

strips of shim stock of appropriate thickness around the edges of the sheet adequately facilitates control of the adhesive thickness. After cure, samples 25.4×178 mm were sawed from the sheet as sketched in Figure 5.16a, after which the lap is formed by making saw cuts as outlined in Figure 5.16b.

After sawing slots to form the lap part of the joint, the resulting specimen appears as shown in the photograph of Figure 5.17. Force can be applied to the specimen either by means of wedge grips or by boring holes near each end to receive a hardened steel pin through which the load is applied. We found the latter to be more satisfactory as it apparently reduces the probability of the grips introducing bending movements to the sample. A U joint in the load string helps to further reduce these moments. The sample is now ready to be pulled to failure. The failure load (and often the deformation) is recorded. Again, because of the viscoelastic–plastic nature of most polymer adhesives, it is essential that the rate of loading be controlled and specified in the reports of test results.

As a final text procedure example, we discuss some peel tests with which we have been involved. We were introduced to these procedures through Dr. Ray Grunzinger of the 3M Company. Dr. Grunzinger with others at 3M developed these tests for investigating the bonding of a cover sheet to a base sheet in reflective sheeting. The procedure could, however, be readily adapted for testing of pressure-sensitive tape and other adhesives used to bond films or their sheets to adherends. We feel these procedures, in general, have much to offer.

For those unfamiliar with the name, *reflective sheeting* is the reflective material on items such as stop signs and freeway direction signs. The largest quantity of this material comprises a polymer backing sheet in which very small partially silvered glass beads have been embedded to a depth of about one-half of their radius. A cover sheet is placed over this, and a network of small interconnecting strips of the cover sheet are heat bonded to the base sheet to form small, completely encapsulated air pockets. Adhesion is extremely important since reflective sheeting is exposed to the weather; should water leak into these pockets, the sign is almost totally blacked out. (This means that all of the reflective qualities of the sign are eliminated; and when a light is shone on it, the sign cannot be read.) Procedures for evaluating the adhesion of the cover sheet to the base sheet are therefore needed, not only for developmental work, but also for quality control on assembly lines.

The tests in this laboratory were conducted by the following procedures. After manufacture (the reflective sheets are in the form of large sheets), strips 2.54 cm (1 in.) wide are then cut from the sheets with the aid of a razor blade tool. Next, a 203-mm-long (8-in.-) strip is bonded to an aluminum plate, as illustrated in Figure 5.18. Both thermally activated adhesive and two-sided pressure-sensitive tape are used for this purpose. A razor blade is then used to initiate a peeled region. For the strength tests, the cover sheet was reinforced with strapping tape to assure that peel failure, rather than cohesive failure of the cover sheet, occurs during loading. After approximately 51 mm (2 in.) of peel have been stripped back with the razor blade, the samples are ready for testing.

To facilitate the peel testing, a test jig is constructed as shown in the photograph of Figure 5.19. The aluminum plate, with the attached sheeting, is loaded into this apparatus by passing it below the two rollers and threading the peel tab of the cover sheet up between the rollers, as shown in Figure 5.19. The end of the tab is gripped by a flat tensile grip. The specimen is peeled by moving the lower crosshead downward at a fixed rate. For most of our tests, this rate is set at 127 mm/min (5 in./min). The output from this test is a record of force versus time (peel displacement). The peel force recording falls into two general classes. Often peel is characterized by the load reaching some peak value followed by rather rapid peel and a drop in load followed by another buildup in load, followed by a subsequent drop. These sawtooth recordings are often called *slip and stick* peel; an example is shown in Figure 5.20. It should be noted that in the specific case of the reflective sheeting, the primary cause of the sawtooth nature of the load-peel curve has to do with the network nature of the bonding. As a

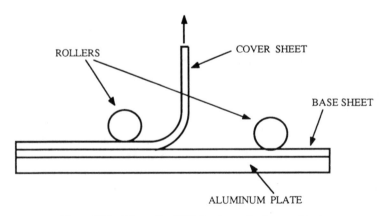

ROLLERS   COVER SHEET

BASE SHEET

ALUMINUM PLATE

Figure 5.18   Schematic of 3M floating roller type peel test.

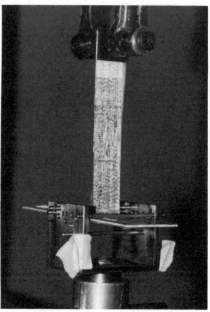

Figure 5.19 Photographs of peel specimen in the 3M floating roller peel apparatus. *Note*: strapping tape on back of cover sheet is to prevent cohesive failure of the cover sheet.

result, the amount of adhesive material being stripped varies as the peel front moves. However, even if every care is taken to make the adhesive as continuous as possible, *slip–stick* behavior is commonly present. There is no complete consensus about the most important part of the results: the peaks, the troughs, or the averages. It is perhaps best to report all three. In other cases, more uniform peel results in a relatively constant load during peel with a constant peel rate (which results in a nearly straight horizontal line on the recorded data). Sometimes, as the peel rate is increased, a transition is observed from smooth peel to slip–stick behavior.

Generally, the results of peel tests are comparative in nature (albeit quantitative). If one wishes to compare peel results between specimens, it is important that: (1) peel rates be the same, (2) the thickness of the peeling adherends be the same, (3) the peel angles be the same, (4) any support films (such as the strapping tape in the example) be identical, and (5) test conditions such as temperature and humidity be carefully controlled.

How one can quantitatively compare test results from one test with those of another type of test is a problem that is not completely resolved; for instance, an adhesive-adherend system that yields high strength in the lap shear test may not exhibit high peel. Fracture mechanics, as outlined earlier, appears to hold the most promise in this respect. We are currently investigating the use of fracture

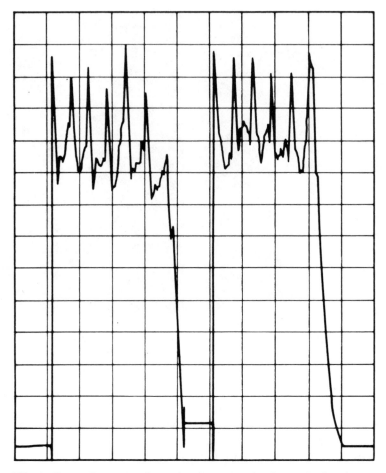

Figure 5.20 An Instron Incorporated recorder chart paper showing results from two peel tests. Time increases from right to left; each major division represents 0.1 min. Vertical axis is load, each major division represents 453.6 kg (1 lb). *Note*: The peaks and valleys in the data in this specific case are in part due to the fact that adhesive is in the form of a grid work, but similar slip-stick type behavior is also often observed for peel in continuous adhesive layers.

mechanics to predict the strength of lap shear and peel specimens from results obtained from tensile adhesion tests. While we have enjoyed noteworthy success with relatively elastic systems, much work remains to be done in this respect, particularly with tougher adhesives that exhibit significant plasticity.

Acknowledgments and Dedication

Much of the research conducted in this area at the University of Utah was supported by or used facilities purchased under National Science Foundation grants; for example, DMR 85-07175. Ms. Johnsen's graduate studies were sponsored by the Utah State Center for Advanced Materials. The efforts of the graduate students are cited as coauthors in our references.

We would like to dedicate this chapter to the memory of Garron P. Anderson, who died of a heart attack in October 1988 in the mountains he loved so much.

References

1. S. R. Hartshorn, Ed., *Structural Adhesives*, Plenum, New York, 1986.
2. R. L. Patrick, Ed., *Treatise on Adhesion and Adhesives*, Vols. 1–6, Dekker, New York, 1966–1988.
3. D. K. Kaelble, *Physical Chemistry of Adhesion*, Wiley, New York, 1971.
4. A. J. Kinloch, *Adhesion and Adhesives*, Chapman & Hall, London, 1987.
5. J. D. Venables, D. K. McNamara, J. M. Chen, T. S. Sun, and R. L. Hopping, *10th National SAMPE Technical Conference Proceedings*, 1978, p. 362.
6. S. S. Voyutskii, Y. I. Marken, V. M. Gorchakova, and V. E. Gul, *Adhes. Age*, **8**, 24 (1965).
7. R. P. Campion, in K. W. Allen, Ed., *Adhesion-1*, Applied Science, London, 1977, p. 63.
8. W. A. Zisman, *Ind. Eng. Chem. Prod. Res. Dev.*, **11**, 170 (1971).
9. E. H. Andrews and A. J. Kinloch, *J. Polym. Sci. Symp.*, **46**, 1 (1974).
10. D. K. Owens, *J. Appl. Polym. Sci.*, **18**, 1869 (1974).
11. I. Kusaka and W. Suetaka, *Spectrochim. Acta*, **36A**, 647 (1980).
12. R. J. Good, in R. L. Patrick, Ed., *Treatise on Adhesion and Adhesives*, Vol. 1, Dekker, New York, 1966, p. 15.
13. A. N. Gent, *Int. J. Adhes. Adhes.*, **1**, 175 (1981).
14. I. E. Klein, J. Sharon, A. E. Yaniv, H. Dodiuk, and D. Katz, *Int. J. Adhes. Adhes.*, **3**, 159 (1983).
15. T. Sugama, L. E. Kukacka, and N. Carciello, *J. Mater. Sci.*, **19**, 4045 (1984).
16. E. P. Plueddemann, *Silane Coupling Agents*, Plenum, New York, 1982.
17. C. H. Chiang, H. Ishida, and J. L. Koenig, *J. Colloid Interface Sci.*, **74**, 396 (1980).
18. M. Gettings and A. J. Kinloch, *J. Mater. Sci.*, **12**, 2049 (1977).
19. B. V. Deryaguin and V. P. Smilga, *Adhesion, Fundamentals and Practice*, McLaren and Son, London, 1969, p. 152.
20. C. Weaver, *Faraday Spec. Discuss. Chem. Soc.*, **2**, 18 (1972).
21. F. M. Fowkes, F. H. Hielscher, and D. J. Kelley, *J. Colloid Sci.*, **32**, 469 (1977).
22. K. L. Mittal, *Pure Appl. Chem.*, **52**, 1295 (1980).
23. L. H. Sharpe and J. H. Schonhorn, in R. F. Gould, Ed., *Advances in Chemistry Series, 43*, American Chemical Society, Washington, DC, 1964, p. 189.
24. R. S. Huntsburger, *Adhes. Age*, **21**, 32, (1978).
25. J. F. Padday, Ed., *Wetting, Spreading and Adhesion*, Academic, New York, 1978.
26. T. Young, *Trans. R. Soc. London*, **95**, 65 (1805).

27. R. J. Good, in D. J. Alner and K. W. Allen, Eds., *Aspects of Adhesion-7*, Transcripta Books, London, 1973, p. 182.

28. W. A. Zisman, in R. F. Gould, Ed., *Advances in Chemistry Series*, *43*, American Chemical Society, Washington, DC, 1964, p. 1.

29. J. R. Dann, *J. Colloid Interface Sci.*, **32**, 302 (1970).

30. R. J. Good and L. A. Girifalco, *J. Phys. Chem.*, **64**, 561 (1960).

31. Y. Tamai, K. Makuuchi, and M. Suzuki, *J. Phys. Chem.*, **71**, 4167 (1967).

32. *The Desk Top Data Bank on Adhesives*, Vols. 1 and 2, Cordura Publications, Inc., LaJolla, CA, 1979.

33. E. P. Pleuddemann, *J. Adhes. Sci. Technol.*, **2**, 179 (1988).

34. ASTM Standards, Volume 15.06, American Society for Testing and Materials, Philadelphia, PA, 1990 (or most current printing).

35. ASTM Standard D897, *Test Method for Tensile Properties of Adhesive Bonds*, Vol. 15.06, American Society for Testing and Materials, Philadelphia, PA, most current printing.

36. G. P. Anderson, S. Chandapeta, and K. L. DeVries, "Effect on Removing Eccentricity from Button Tensile Tests," in S. Johnson, Ed., *Adhesively Bonded Joints: Testing, Analysis, and Design*, ASTM, STP-981, American Society of Testing and Materials, Philadelphia, PA, 1988.

37. ASTM Standard D2094, *Recommended Practice for Preparation of Bar and Rod Specimens for Adhesion Tests*, Vol. 15.06, American Society for Testing and Materials, Philadelphia, PA, most current printing.

38. ASTM Standard C297, *Method for Tension Test of Flat Sandwich Constructions in Flatwise Plane*, Vol. 15.03, American Society for Testing and Materials, Philadelphia, PA, most current printing.

39. ASTM Standard D1344, *Methods for Testing Cross-Lap Specimens for Tensile Properties of Adhesives*, Vol. 15.06, American Society for Testing and Materials, Philadelphia, PA, 1978 (or most current printing).

40. G. P. Anderson and K. L. DeVries, "Analysis of Standard Bond Strength Tests," in R. Patrick, K. L. DeVries, and G. P. Anderson Eds., *Adhesion and Adhesives*, Vol. 6, Dekker, New York, 1988.

41. ASTM Standard D1876, *Test Method for Peel Resistance of Adhesives (T-Peel Test)*, Vol. 15.06, American Society for Testing and Materials, Philadelphia, PA, most current printing.

42. ASTM Standard D3167, *Test Method for Floating Roller Peel Resistance of Adhesives*, Vol. 15.06, American Society for Testing and Materials, Philadelphia, PA, most current printing.

43. ASTM Standard D1791, *Test Method for Accelerated Aging of Liquid Water-Emulsion Floor Polishes*, Vol. 15.04, American Society for Testing and Materials, Philadelphia, PA, 1982 (or most current printing).

44. ASTM Standard D903, *Test Method for Peel of Stripping Strength of Adhesive Bonds*, Vol. 15.06, American Society for Testing and Materials, Philadelphia, PA, most current printing.

45. G. R. Hamed, "Energy Conservation During Peel Testing," in R. L. Patrick, K. L. DeVries, and G. P. Anderson Eds., *Adhesion and Adhesives*, Vol. 6, Dekker, New York, 1988.

46. A. N. Gent and G. R. Hamed, *J. Appl. Polym. Sci.*, **21**, 2817 (1977).

47. A. N. Gent and G. R. Hamed, *Polym. Eng. Sci.*, **17**, 462 (1977).

48. ASTM Standard D1002, *Test Method for Strength Properties of Adhesives in Shear by Tension Loading (Metal-to-Metal)*, Vol. 15.06, American Society for Testing and Materials, Philadelphia, PA, most current printing.

49. ASTM Standard D3165, *Test Method for Strength Properties of Adhesives in Shear by Tension Loading of Laminated Assemblies*, Vol. 15.06, American Society for Testing and Materials, Philadelphia, PA, most current printing.

50. ASTM Standard D3528, *Test Method for Strength Properties of Double Lap Shear Adhesive Joints by Tension Loading*, Vol. 15.06, American for Testing and Materials, Philadelphia, PA, most current printing.

51. S. S. Wang and J. F. Yau, *Int. J. Fract.*, **19**, 295 (1982).

52. ASTM Standard D3363, *Test Method for Film Hardness by Tensile Test*, Vol. 6.01, American Society for Testing and Materials, Philadelphia, PA, 1974 (or most current printing).

53. J. C. McMillian, "Surface Preparation—The Key to Bondment Durability," *Bonded Joints and Preparation for Bonding—AGARD Lecture Series No. 102*, Technical Editing and Reproduction Ltd., London, 1979, Chap. 7.

54. ASTM Standard D2293, *Test Method for Creep Properties of Adhesives in Shear by Compression Loading (Metal-to-Metal)*, Vol. 15.06, American Society for Testing and Materials, Philadelphia, PA, most current printing.

55. ASTM Standard D2294, *Test Method for Creep Properties of Adhesives in Shear by Tension Loading (Metal-to-Metal)*, Vol. 15.06, American Society for Testing and Materials, Philadelphia, PA, most current printing.

56. ASTM Standard D1780, *Recommended Practice for Conducting Creep Tests of Metal-to-Metal Adhesives*, Vol. 15.06, American Society for Testing and Materials, Philadelphia, PA, most current printing.

57. ASTM Standard D2295, *Test Method for Strength Properties of Adhesives in Shear by Tension Loading at Elevated Temperatures (Metal-to-Metal)*, Vol. 15.06, American Society for Testing and Materials, Philadelphia, PA, most current printing.

58. ASTM Standard D2257, *Strength and Properties of Adhesives in Shear by Tension Loading in the Temperature Range From -267.8 to $-55°C$*, Vol. 15.06, American Society for Testing and Materials, Philadelphia, PA, most current printing.

59. ASTM Standard D1382, *Test Method for Susceptibility of Dry Adhesive Films to Attack by Roaches*, Vol. 15.06, American Society for Testing and Materials, Philadelphia, PA, most current printing.

60. ASTM Standard D1383, *Test Method for Susceptibility of Dry Adhesive Films to Attack by Laboratory Rats*, Vol. 15.06, American Society for Testing and Materials, Philadelphia, PA, most current printing.

61. ASTM Standard D3434, *Recommended Practice for Multiple-Cycle Accelerated Age Test (Automatic Boil Test) for Exterior Wet Use Wood Adhesives*, Vol. 15.06, American Society for Testing and Materials, Philadelphia, PA, most current printing.

62. ASTM Standard D3632, *Practice for Accelerated Aging of Adhesive Joints by the Oxygen-Pressure Method*, Vol. 15.06, American Society for Testing and Materials, Philadelphia, PA, most current printing.

63. ASTM Standard D2651, *Recommended Practice for Preparation of Metal Surfaces for Adhesive Bonding*, Vol. 15.06, American Society for Testing and Materials, Philadelphia, PA, most current printing.

64. ASTM Standard D3933, *Practice for Preparation of Aluminum Surfaces for Structural Adhesives Bonding*, Vol. 15.06, American Society for Testing and Materials, Philadelphia, PA, most current printing.

65. ASTM Standard D2093, *Recommended Practice for Preparation of Surfaces of Plastics Prior to Adhesive Bonding*, Vol. 15.06, American Society for Testing and Materials, Philadelphia, PA, most current printing.

66. A. A. Griffith, 1st Proceedings of the International Congress of Applied Mechanics, Delft, Holland, 1924, p. 55.

67. M. L. Williams, *Bull. Seismol. Soc. Am.*, **49(2)**, 199 (1959).

68. E. H. Andrews, T. A. Khan, and H. A. Majid, *J. Mater. Sci.*, **20**, 3621 (1985).

69. E. H. Andrews, H. A. Majid, and N. A. Lockington, *J. Mater. Sci.*, **19**, 73 (1984).

70. D. W. Aubrey and M. Sherriff, *J. Polym. Sci. Polym. Chem. Ed.*, **18**, 2597 (1980).

71. W. D. Bascom and J. Oroshnik, *J. Mater. Sci.*, **10**, 1411 (1978).

72. W. D. Bascom and D. L. Hunston, "Fracture of Epoxy and Elastomer-Modified Epoxy Polymers," in R. L. Patrick, K. L. DeVries, and G. P. Anderson, Eds., *Treatise on Adhesion and Adhesives*, Vol. 6, Dekker, New York, 1988, p. 123.

73. H. F. Brinson, J. P. Wightman, and T. C. Ward, *Adhesives Science Review 1*, VPI Press, Blacksburg, VA, 1987.

74. D. Broek, *Elementary Engineering Fracture Mechanics*, Noordoff, Leyden, 1974.

75. J. D. Burton, W. B. Jones, and M. L. Williams, *Trans. Soc. Rheol.*, **15**, 39 (1971).

76. G. Danneberg, *J. Appl. Polym. Sci.*, **5**, 125 (1961).

77. F. Erdogan, *Eng. Fract. Mech.*, **4**, 811 (1972).

78. A. N. Gent, *Int. J. Adhes. Adhes.*, **1**, 175 (1981).

79. A. N. Gent and G. R. Hamed, *Plast. Rubber Proc.*, **3**, 17 (1977).

80. T. R. Guess, R. E. Allred, and F. P. Gerstle, *J. Test. Eval.*, **5(2)**, 84 (1977).

81. G. R. Hamed, "Energy Conservation During Peel Testing," in R. L. Patrick, K. L. DeVries, and G. P. Anderson, Eds., *Treatise on Adhesion and Adhesives*, Vol. 6, Dekker, New York, 1988, p. 33.

82. R. W. Hertzberg and J. A. Manson, *Fatigue of Engineering Plastics*, Academy, Santz Cruz, CA, 1980.

83. G. R. Irwin, "Fracture Mechanics Applied to Adhesive Systems," in R. L. Patrick, Ed., *Treatise on Adhesion and Adhesives*, Vol. 1, Dekker, New York, 1966, p. 233.

84. W. S. Johnson, *J. Test. Eval.*, **15(6)**, 303 (1987).

85. D. H. Kaelble, *Physical Chemistry of Adhesion*, Wiley-Interscience, New York, 1971.

86. H. H. Kaush, *Polymer Fracture*, Springer-Verlag, Berlin, 1978.

87. A. J. Kinloch, in A. J. Kinloch, Ed., *Structural Adhesives: Developments in Resins and Primers*, Applied Science, London, 1986, p. 127.

88. W. G. Knauss and K. M. Liechti, "Interfacial Crack Growth and Its Relation to Crack Front Profiles," *ACS Organic Coatings and Applied Polymer Science Proceedings*, Vol. 47, American Chemical Society, Wasington, DC, 1982, p. 418.

89. K. M. Liechti and C. Lin, *Structural Adhesives in Engineering*, Institute of Mechanical Engineering, London, 1986, p. 83.

90. J. C. McMillian, *Developments in Adhesives*, 2nd ed., Applied Science, London, 1981, p. 243.

91. S. Mostovoy and E. J. Ripling, *Adhes. Sci., Technol.*, **9B**, 513 (1975).

92. D. R. Mulville, D. L. Hunston, and P. W. Mast, *J. Eng. Mater. Technol.*, **100**, 25 (1978).

93. J. R. Rice and G. C. Sih, *J. Appl. Mech.*, **32**, 418 (1965).

94. E. J. Ripling, J. S. Santner, and P. B. Crosley, *J. Mater. Sci.*, **18**, 2274 (1983).

95. E. F. Rybicki and M. F. Kanninen, *Eng. Fract. Mech.*, **9**, 921 (1974).

96. G. B. Sinclair, *Int. J. Fract.*, **16**, 111 (1980).

97. J. D. Venables, D. K. McNamara, J. M. Chen, T. S. Sun, and R. L. Hopping, *Appl. Surf. Sci.*, **3**, 88 (1979).

98. S. S. Wang, J. F. Mandell, and F. J. McGarry, *Int. J. Fract.*, **14**, 39 (1978).

99. S. S. Wang and J. F. Yau, *Int. J. Fract.*, **19**, 295 (1982).

100. J. G. Williams, *Fracture Mechanics of Polymers*, Ellis Horwood, Chichester, 1984.

101. M. L. Williams, *J. Adhes.*, **5**, 81 (1973).

102. R. J. Young, in A. J. Kinloch, Ed., *Structural Adhesives: Developments in Resins and Primers*, Applied Science, London, 1986, p. 163.

103. V. L. Hein and F. Erdogan, *Int. J. of Fract. Mech.*, **7** 317 (1971).

104. F. Delale, F. Erdogan, and M. M. Aydinoglu, *J. Compos. Mater.*, **15**, 1123 (1981).

105. G. P. Anderson and K. L. DeVries, *J. Adhes.*, **23**, 289 (1987).

106. G. P. Anderson, S. J. Bennett, and K. L. DeVries, *Analysis and Testing of Adhesive Bonds*, Academic, New York, 1977.

107. J. K. Strozier, K. J. Ninow, K. L. DeVries, and G. P. Anderson, *Adhes. Sci. Rev.*, **1**, 121 (1987).

108. G. P. Anderson, D. H. Brinton, K. J. Ninow, and K. L. DeVries, "A Fracture Mechanics Approach to Predicting Bond Strength," in S. Mall, K. M. Liechti, and J. K. Vinson, Eds., *Advances in Adhesively Bonded Joints, Proceedings of a Conference at the Winter Annual Meeting of ASME*, November 27-December 2, 1988, Chicago, IL, ASME, New York, 1988, pp. 98–101.

109. ASTM Standard D3433, *Test Method for Fracture Strength in Cleavage of Adhesives in Bonded Joints*, Vol. 15.06, American Society for Testing and Materials, Philadelphia, PA, most current printing.

Chapter **6**

DETERMINATION OF THE BULK MODULUS AND COMPRESSIVE PROPERTIES OF SOLIDS

Bruce Hartmann

1   INTRODUCTION

1.1   Overview

Under an applied pressure, the volume of a solid decreases. This is a simple consequence of LeChatelier's principle or, equivalently, of thermodynamic stability for a reversible system. The bulk modulus is a measure of the change in

267

pressure required to produce a given decrease in volume, and it is defined as

$$K_T = -V(\partial P/\partial V)_T \qquad (1)$$

where V is volume, P is pressure, and T is absolute temperature. Since the derivative is evaluated at constant temperature, (1) yields the isothermal bulk modulus. The subscript is often omitted and the quantity is simply referred to as the bulk modulus K. Note that the minus sign appears so that K is a positive quantity, since the derivative is negative. As seen in (1), bulk modulus has dimensions of pressure and is expressed here in units of bars: 1 bar $= 10^6$ dyn/cm^2 $= 10^5$ Pa $= 0.9869$ atm $= 1.020$ kg/cm^2.

Illustrative values of bulk modulus for several solids at room temperature and pressure are compiled from various sources [1–9] in Table 6.1. (Two liquids, mercury and water, are included for the sake of comparison.) The values vary by a factor of 400, with diamond the highest and cesium the lowest. Also listed in Table 6.1 (where available) are the rates of change of bulk modulus with

Table 6.1 Typical Bulk Modulus Values[a,b]

Material	K (kbar)	dK/dP	$-(dK/dT)$ (kbar/K)
Diamond	5550		0.18
Tungsten	3130		
Platinum	2790		
Iron	1680	4.7	
Copper	1370	5.5	
Silicon	990	4.9	
Aluminum	730	5.3	
Lead	430	5.1	
Quartz	370	6.5	
Mercury	260	9.2	0.28
Granite	120		
Poly(methyl methacrylate)	65	13	0.13
Nylon-6,6	65		
Poly(ethylene)	45	8	0.23
Poly(styrene)	42	11	0.07
Poly(propylene)	44	18	0.44
Poly(tetrafluoroethylene)	21	16	
Poly(ethylene oxide)	56	10	0.91
Water	20		
Cesium	14		

[a]At room temperature and atmospheric pressure, determined using various techniques. These values should be considered illustrative and not necessarily used for design purposes.
[b]Data taken from [1–9].

temperature and pressure. The values in Table 6.1 should be considered representative rather than unique. Different experimental techniques give somewhat different results; the values generally vary with the purity, morphology, and thermal history of the sample.

Compressive properties other than bulk modulus can also be determined from pressure measurements. In some cases, the pressure dependence of the volume of a solid shows a more or less abrupt change in behavior, which is indicative of a phase change. There are two principal classes of thermodynamic phase changes or transitions: first-order transitions, where there is a change in volume at the transition, and second-order transitions, where the volume is unaltered but there is a change in the derivative of the volume. Melting and crystal–crystal transitions are first order, while the glass transition is second order. Thus, by performing volume measurements over a range of pressure and temperature, the phase diagram of a solid can be determined.

A phase formed under pressure may be stable at atmospheric pressure. A particular example that has intrigued high-pressure workers for many years is the formation of diamond from graphite. Although the phase diagram shows that diamond can be formed at 15 kbar at room temperature, the conversion proceeds at a very sluggish rate. The first report on the formation of synthetic diamonds [10] in 1955 made use of 100 kbar at 2000 K for many hours. Today, synthetic commercial diamonds are made routinely by the use of high pressure.

At low pressure, in the absence of phase transitions, the effect of pressure on a solid is to compress the free volume between molecules and to distort the intermolecular distances and bond angles, but the atoms are unaltered structurally. At higher pressure (ca. 100 kbar), individual atoms are "crushed" and electron shells are rearranged. Theoretically, at high enough pressure all solids should behave like metals as a result of this rearrangement. The required pressure is a few megabars for hydrogen.

The general types of apparatus that are used to measure the bulk modulus and compressive properties of solids are idealized in Figure 6.1. Figure 6.1a shows an idealized application of true hydrostatic pressure. As the word hydrostatic implies, the stress (force per unit area) is applied to a fluid and acts equally in all directions. Thus, the action of hydrostatic pressure is to produce a change in volume with no change in shape. (By contrast, a shear stress produces a change in shape with no change in volume.) Measurements made as in Figure 6.1a are the most accurate, but the requirement for a liquid to transmit the pressure to the sample limits the pressure range. An apparatus of the type shown in Figure 6.1a is called a *dilatometer*. Figure 6.1b shows a simple apparatus that can be used up to higher pressure. Here, the stress is applied to the sample along only one direction. The sides try to move outward in response to the stress, but they are restrained by the rigid container. Stresses are thus generated on the sides as well as on the top and bottom. Since the stresses on the sides are less than on the top and bottom, the stress field is not truly hydrostatic, but it is approximately so and is often called *quasihydrostatic*. The limitation of this type of apparatus is the yield strength of the container. Figure 6.1b illustrates a

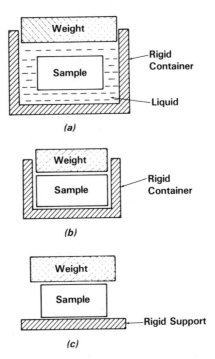

Figure 6.1 Idealized methods of applying compression: (a) true hydrostatic compression, (b) quasihydrostatic compression, and (c) uniaxial compression.

piston–cylinder apparatus. In Figure 6.1c a uniaxial load is applied to a sample whose sides are unconstrained. The stress field is not truly hydrostatic, but this technique is sometimes the only way to achieve very high stresses.

This chapter begins with a discussion of the necessary background to prepare the reader for the experimental techniques that are emphasized. In the remainder of this chapter we consider five different experimental techniques: dilatometer, piston–cylinder, acoustics, diamond anvil cell, and shock waves. For each technique, the apparatus is described, usually using a particular device as an illustration, and then representative results obtained with that apparatus are presented. Results for polymeric solids are emphasized because they display the complexities of these measurements (pressure, temperature, and time dependence, as well as phase transitions) at easily reached pressure and temperature.

1.2   Background

As seen in (1), we must evaluate the derivative of pressure versus volume data to determine K_T. The same experimental data can also be analyzed to evaluate the inverse derivative

$$k_T = -(1/V)(\partial V/\partial P)_T \tag{2}$$

which is known as the *isothermal compressibility*. Since k_T is simply the reciprocal of K_T,

$$k_T = 1/K_T \qquad (3)$$

it is a matter of convenience whether the data is analyzed in terms of bulk modulus or compressibility. Knowing one is the same as knowing the other. They are equivalent measures of the same physical property: the response of volume to pressure changes.

Volume depends on temperature as well as on pressure. The temperature dependence of volume is expressed by the thermal expansion coefficient, which is defined as

$$\alpha = (1/V)(\partial V/\partial T)_P \qquad (4)$$

so that k_T and α are complementary quantities in principle and experimentally they are often determined in the same apparatus and can be calculated from the same pressure–volume–temperature (PVT) data.

To determine K_T from (1) or k_T from (2), a derivative must be evaluated. Experimental volume isotherms can be differentiated numerically, but it should be realized that this differentiation is one of the major sources of error in the determination of the bulk modulus. Alternatively, the experimental data can be fitted to an analytic equation for volume, which can then be differentiated exactly, but the validity of the analytic equation can be questioned. Both approaches are outlined here.

A convenient method of numerically differentiating experimental data is a moving five-point quadratic least-square fitting procedure. Approximating (1) with finite differences yields

$$K \approx -V(\Delta P/\Delta V) \qquad (5)$$

where ΔP is the pressure interval for which the volume change is ΔV. It is convenient to define volume strain, also called *compression*, as

$$\varepsilon = -(V - V_0)/V_0 \qquad (6)$$

where V_0 is the volume at the initial (nominally zero) pressure. Then at five equally spaced strains (labeled -2, -1, 0, 1, and 2) the value of the pressure is recorded. A quadratic curve is fitted to these points using the method of least squares, and the derivative of the resulting curve is evaluated at the midpoint (labeled 0) using the formula [11]

$$K_T(0) = [-2P(-2) - P(-1) + P(1) + P(2)]/10\Delta\varepsilon \qquad (7)$$

where the numbers in parentheses refer to the five strain labels and $\Delta\varepsilon$ is the spacing between the strains. The procedure is repeated throughout the pressure range. As a rule of thumb, numerical differentiation magnifies experimental uncertainty by a factor of about 10. Thus, if volume is measured to an accuracy of 0.1%, then bulk modulus is accurate to 1%.

The other approach for evaluating K_T is to use an analytic equation called a *PVT equation of state*, which is an equation for the pressure and temperature dependence of the volume, $V(T, P)$. (Even though bulk modulus involves only a pressure derivative, temperature is included as a variable to determine the temperature dependence of the bulk modulus.) One starting point in deriving an equation of state is to note that the physical effects of pressure and temperature usually oppose each other; that is, as pressure increases volume decreases, and as temperature increases volume increases. This observation suggests that bulk modulus is a function of volume only. Then, the dependence of bulk modulus on temperature and pressure arises because volume depends on temperature and pressure. This behavior results if the bulk modulus depends only on the intermolecular distance between molecules. Equations of state derived from this model have been presented both for low molecular weight solids [12] and for polymers [13–15]. In polymers, the model takes into account that compression occurs primarily in only two dimensions because the covalent bonding along the chains is much stiffer than the van der Waals bonding between the chains.

An equation of state allows the calculation not only of the modulus, but also of the volume derivative of the bulk modulus, which is a fundamental quantity from solid state physics known as the *Gruneisen parameter*

$$\gamma = -\tfrac{1}{2}(d\ln K/d\ln V) \tag{8}$$

The derivative is usually evaluated experimentally in one of two ways: from the temperature dependence of the volume,

$$\gamma = -\tfrac{1}{2}(\partial \ln K/\partial T)_P/(\partial \ln V/\partial T)_P$$
$$= -\tfrac{1}{2}\alpha(\partial \ln K/\partial T)_P \tag{9}$$

or from the pressure dependence,

$$\gamma = -\tfrac{1}{2}(\partial \ln K/\partial P)_T/(\partial \ln V/\partial P)_T$$
$$= \tfrac{1}{2}(\partial K/\partial P)_T \tag{10}$$

From the values of dK/dP listed in Table 6.1, it is seen that the Gruneisen parameter for polymers typically lies between 4 and 8, while for metals the value is 2–3. Several studies [7, 13–17] examined the Gruneisen parameter for polymers in more detail.

Other equations of state were derived from a variety of different empirical and theoretical starting points. Among polymer equations of state are the empirical Tait equation (to be displayed later), the hole model of Simha–Somcynsky [18], and a simpler semiempirical model that is applied to both solid and liquid polymers [19]. Generally, equations of state are able to reproduce the experimental volume measurements fairly well, at least up to about 2-kbar pressure.

Bulk modulus is not only a thermodynamic property, depending on the variables P, V, and T, but it is also one of the constants variously called an *elastic stiffness constant, elastic constant, modulus of elasticity*, or *modulus*. In general, a modulus is the ratio of a stress to a strain. For bulk modulus, the stress field is that of hydrostatic pressure, but other stress fields and other elastic constants are possible. For example, the ratio of shear stress to shear strain is called the *shear modulus G*, and the ratio of uniaxial stress (either tension or compression) to strain is called the *Young's modulus E*. A fourth constant often used is the ratio of fractional lateral contraction to fractional uniaxial extension, called *Poisson's ratio σ*. For an isotropic solid, only two of these constants are independent. The others can be calculated when any two are known. In general, the reciprocal of a modulus is called a *compliance*. Thus, compressibility is the compliance reciprocal to the bulk modulus.

In terms of these elastic constants, the modulus in Figure 6.1a is K, while for Figure 6.1b the modulus is [20]

$$M = K + 4G/3 \tag{11}$$

so that measurements made in this manner are accurate if G is much less than K or if a correction for G can be made. Finally, for Figure 6.1c, the modulus is

$$E = 9KG/(3K + G) \tag{12}$$

and generally shows the greatest difference from K.

The four constants K, G, E, and σ are usually referred to as *engineering moduli*. Equivalent elastic constants can also be defined from the generalized tensor form of Hooke's law, which states that stress is directly proportional to strain

$$\sigma_i = c_{ij}\varepsilon_j \tag{13}$$

where σ_i is stress (force per unit area) in the i direction and ε_j is strain (change in length per unit initial length) in the j direction. The constants of proportionality c_{ij} are also called *elastic stiffness constants* or *modulus values*. The set of values c_{ij} is a 6×6 matrix where, because of symmetry requirements, not all elements are independent. In the most unsymmetric crystal (triclinic), there are 21 independent elastic constants. For a cubic crystal, there are three; for an isotropic solid, there are only two independent elastic constants, c_{11} and c_{12}. These values are related to K and G by the relations [20]

$$K = \tfrac{1}{3}(c_{11} + 2c_{12}) \tag{14}$$

$$G = \tfrac{1}{2}(c_{11} - c_{12}) \tag{15}$$

Another equivalent set of elastic constants are the Lame constants λ and μ, which are related to engineering constants by the relations

$$K = \lambda + \tfrac{2}{3}\mu \tag{16}$$

$$G = \mu \tag{17}$$

The different sets of elastic constants express the same physical property, and it is a matter of preference which system to use (at least for isotropic solids). Since the different systems are used in the literature, one must be familiar with their interrelations.

No discussion of pressure measurements is complete without mentioning P. W. Bridgman, who was with the Physics Department of Harvard University from 1908 to 1961. His life was devoted to pressure measurements, which he did with relatively simple equipment, and he almost always worked alone. Of his more than 100 papers, only one has a coauthor. In 1946, he was awarded the Nobel prize in physics. His influence is pervasive and much of this presentation is derived from his work. A useful summary of his contributions [21] and the state of the art of pressure measurements in 1946 includes 674 references.

2   DILATOMETER

2.1   Description of Apparatus

For applications in which hydrostatic pressures up to about 2 kbar are required, a dilatometer is often used. In principle the method is the same as a mercury-in-glass thermometer. The sample is placed in a glass tube, and a confining liquid, usually mercury, is introduced into the tube so that liquid surrounds the sample and extends partway up a glass capillary tube. A capillary tube is used so that relatively small changes in volume produce easily measured changes in the height of the mercury in the capillary. The steps in the construction of the dilatometer [22] are shown in Figure 6.2. The tubing from which the bulb of the dilatometer is made can be any size convenient to fit the size and shape of the sample; 20-mm-diameter tubing is typical. The ability to use irregularly shaped samples and even several pieces of samples is one of the advantages of a dilatometer.

The size of the capillary depends on the volume change expected, which depends on the sample properties, sample size, and range of pressure and temperature used. In a typical example [22], the capillary is Pyrex glass, 500 mm long, 2 mm inner diameter, and 3 mm wall thickness. The capillary is specially selected for uniformity of bore and is graduated along its length in millimeters. Note that the greater the uniformity of bore throughout the length of the capillary, the simpler are the calculations and more reliable are the results.

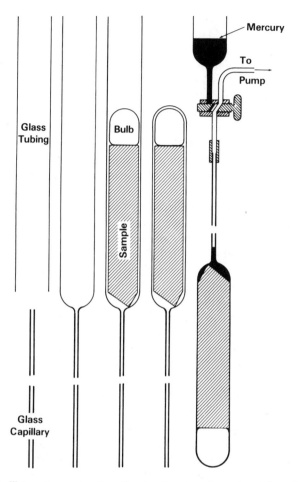

Figure 6.2 Basic dilatometer construction. Construction steps shown from left to right. Reprinted from *Journal of Research of the National Bureau of Standards*; N. Bekkedahl, *J. Res. Natl. Bur. Stand.*, **42**, 145 (1949).

The first step in using a dilatometer is to calibrate the capillary tube by measuring the weight of mercury that corresponds to measured changes in height. The calibration is performed as follows. After filling the capillary, release about 10% of the height of the mercury, measuring to the nearest 0.5 mm, and determine the weight of the mercury to the nearest 10 mg. The volume of the capillary between these two levels is determined from the weight of mercury and its density. This procedure is repeated for the length of the capillary. From these measurements, an overall average value for the capillary is found and also the deviations from the average at each location. For a 2-mm-diameter capillary, a typical average value [22] is 0.003720 cm^3/mm. There are commercial precision bore capillary tubes that are very uniform, but this fact should be verified before using a single number to characterize the tube. After calibrating the capillary tube, it is attached to the dilatometer tube by standard glassblowing procedures.

After weighing the sample to the nearest milligram, it is inserted into the dilatometer. A hollow bulb of the same type of glass as the dilatometer is added, and the end is sealed off to form the completed dilatometer. The purpose of the bulb is to allow the sealing of the dilatometer without unduly heating the sample and without increasing the net volume of mercury in the dilatometer, which would decrease the precision of the measurements. To avoid the glassblowing step of sealing the end, a ground-glass stopper can be used, but this technique is not recommended. When the sample is soft enough to flow, it may clog the capillary. The capillary can then be bent near the bottom into a U shape, thus inverting the bulb.

At this point, the system is filled with the confining liquid, which must not swell or otherwise interact with the sample, remain liquid over the entire range of use, and have known compressibility and thermal expansion. Mercury is almost always the liquid of choice. Filling is done under vacuum to prevent air from being trapped in the system, which is one of the commonest problems with this type of measurement. Not shown in Figure 6.2 is a fine wire, 0.4 mm diameter, that reaches to the bulb and rests on the sample. (The wire must be inert to the sample and to mercury.) The object of the wire is to allow any trapped air to pass out more freely. Then vacuum is applied, possibly for as along as several days, to remove gas dissolved in the sample. This step is critical because the gas properties are so different from the solid sample properties. The wire is then removed.

The completed dilatometer is placed in a liquid thermostat bath regulated to $\pm 0.05°C$. Occasionally, a continuously increasing temperature ramp is used, but stepwise isothermal measurements are more accurate since any thermal lag effects are eliminated. Accurate volume measurements require a stem correction [22] for those parts of the thermometer (if used) and capillary that extend above the level of the heat bath liquid and hence are not at the same temperature as the rest of the system. To make this correction as well as to calibrate the capillary tube and to make later calculations, the volume of mercury as a function of temperature and pressure is needed. These values are available [2] for pressure to 13 kbar and temperature from 22 to 53°C and also [23] for pressure to 8 kbar and temperature from 30 to 150°C.

The dilatometer measurement yields the change in volume relative to the absolute value at room temperature and pressure, and volume measurements can be no more accurate than this value. If the absolute value is measured to $\pm 2 \times 10^{-4}$ cm^3/g, all other measurements should be made to maintain this level of accuracy. For the apparatus described here, the relative error in volume is $\pm 3 \times 10^{-4}$. The precision can be increased by more accurately measuring and controlling temperature and by using a smaller diameter capillary, although then only a smaller total volume change can be measured.

Pressure control must be added to the foregoing temperature control system. A typical example [24] is shown in Figure 6.3. The capillary is open at the top,

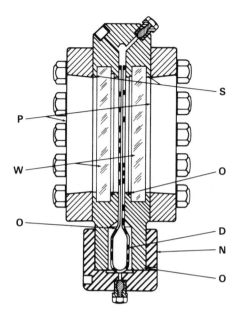

Figure 6.3 Pressurized dilatometer: S, metal spacers; O, O rings; P, cover plate; D, dilatometer; W, glass windows; N, large nut. Reprinted with permission from G. M. Martin and L. Mandelkern, *J. Appl. Phys.*, **34**, 2312 (1963).

and hydrostatic pressure is applied both inside and outside the dilatometer. The pressure system contains a lead screw turned by hand, which advances a piston in a steel chamber and can produce pressures up to 1 kbar. The pressure generator is connected by metal pressure tubing to a blowout disk, a large Bourdon gauge, and a steel pressure vessel immersed in a constant temperature bath. The pressure-transmitting liquid is a diester, di-(2-ethylhexyl)sebacate. This liquid can be used for temperatures from -30 to $150°C$. A cathetometer is used to observe, through the glass windows in the pressure vessel, the mercury level in the capillary.

A variation of the standard dilatometer technique is to use a bellows arrangement to detect the volume change of the mercury. In this technique [25, 26] the sample and confining liquid are sealed in a flexible metal bellows, which is then subjected to hydrostatic pressure. Volume change is obtained from measurements of the change in length of the bellows. A simplified diagram of the apparatus [27] is shown in Figure 6.4. The sample and confining liquid are enclosed in a rigid sample cell, one end of which is closed by a flexible metal bellows. Hydrostatic pressure is applied to the outside of this cell using silicone oil as the pressure-transmitting liquid. The pressure is transmitted to the contents by the flexible bellows, which expands until the pressure in the cell equals the applied pressure, except for a small pressure difference resulting from the spring constant of the bellows (amounting to less than 1 bar). The length change of the bellows is a measure of the volume change of the contents, provided the effective cross-sectional area of the bellows is a known constant. The length change is measured by a linear variable differential transformer

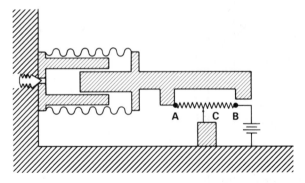

Figure 6.4 Simplified diagram of bellows apparatus: AB, slide wire; C, contact. Reprinted with permission from W. A. Steele and W. Webb, "Compressibility of Liquids," in R. S. Bradley, Ed., *High Pressure Physics and Chemistry*, Vol. 1, Academic, New York, 1963, pp. 145–176.

(LVDT), the coil of which is mounted outside the pressure vessel. The design uses the off-null voltage of the LVDT directly to display length changes on a digital indicator. The sample cell is made from stainless steel, and it is filled with mercury under a vacuum of 50 μbar. Pressures of 2.2 kbar are reached using a hand pump. Pressure measurements are made by using a precision Bourdon pressure gauge. The first step with this apparatus is a calibration run in which the chamber is filled with mercury only. Specific volume changes can be measured with an error of ± 0.0015 cm^3/g or less. The temperature range of the apparatus is from 30 to 350°C. An alternate source of pressurization is a standard nitrogen compressed gas cylinder [28]. When using compressed gas as the pressure source, some gas may dissolve in the polymer. The dissolved gas can have a plasticizing effect, which is often significant in crazing, but is not expected to affect modulus measurements.

Another method of measuring the change in the mercury height resulting from a volume change is through the use of sound waves [29]. The apparatus is shown in Figure 6.5. The device consists of two Invar sections fastened together with Invar screws and sealed with O rings. The use of Invar minimizes the thermal expansion of the container. The upper chamber contains a 23.7-cm^3 sample chamber and a precision bore 6.35 cm high and 1.110 cm in diameter. The sample chamber and bore are connected by a small diameter passageway. The lower section contains a 5-MHz quartz transducer placed at the bottom of the bore. The material under study is placed in the sample chamber, which is then filled under vacuum with mercury. A pulse of sound is emitted by the quartz transducer and passes up the mercury column to be reflected at the mercury–float interface and returned to the transducer. The total time of flight is measured and, knowing the speed of sound in mercury, we can calculate the distance between transducer and mercury-float interface. Pressure changes are made by placing the device in a conventional pressure chamber, with a 1-kbar capability. A floating deadweight pressure balance is used for the pressure

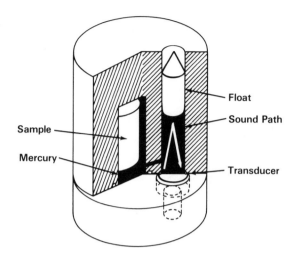

Figure 6.5   Acoustic dilatometer. Reprinted with permission from R. D. Corsaro, J. Jarzynski, and C. M. Davis, Jr., *J. Appl. Phys.*, **45**, 1 (1974).

measurement. The entire pressure vessel is submerged in a large oil bath containing heaters. The system maintains the temperature of the bath to within 1 mK.

2.2   Representative Results

As an example of the data generated in a bellows apparatus, Figure 6.6 shows values obtained [30] for PMMA. To determine the bulk modulus, the data was fitted to the empirical Tait equation. For this equation of state, the bulk modulus is given by

$$K_T(\text{Tait}) = [B(T) + P]\{11.19 - \ln[1 + P/B(T)]\} \tag{18}$$

where

$$B(T) = B_0 \exp(-B_1 T) \tag{19}$$

For the data shown in Figure 6.6, $B_0 = 3564$ bar and $B_1 = 3.229 \times 10^{-3}$ K^{-1}, so that $K_T = 37$ kbar at room temperature and pressure. From (18), the rates of change of K_T with temperature and pressure can be calculated to yield $dK_T/dT = -0.12$ kbar/K and $dK_T/dP = 10$.

Typical results from a standard dilatometer [31] are shown in Figure 6.7 for poly(propylene). (For ease in plotting, an arbitrary straight line was subtracted from the observed data. The volume units in Figure 6.7 are thus arbitrary.) As can be seen, there is a break in the isobars at a temperature that increases with pressure. This break is caused by the second-order glass transition and is a

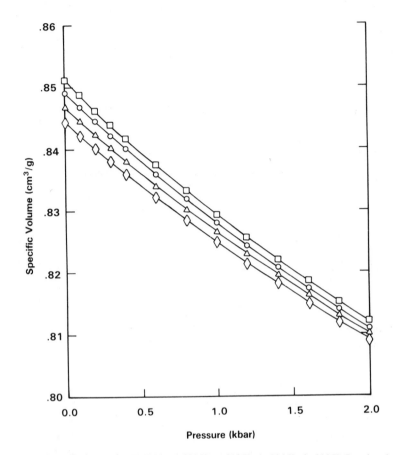

Figure 6.6   Volume isotherms for PMMA: □, 330 K; ○, 319 K; △, 305 K; ◇, 290 K. Reprinted with permission from O. Olabisi and R. Simha, *Macromolecules*, **8**, 206 (1975). Copyright 1975 American Chemical Society.

function of pressure [32]. The pressure dependence is shown in Figure 6.8. The slope of this plot is 20 K/kbar, which is typical for the glass transition of a polymer.

3   PISTON–CYLINDER DEVICE

3.1   Description of Apparatus

The pressure in this device is quasihydrostatic and is generated as shown in Figure 6.1*b*. Warfield's version of the method is described. He used this technique in a series of papers on the bulk modulus and compressive properties of solid polymers [4, 5, 7, 33–37], and the technique is summarized in [38].

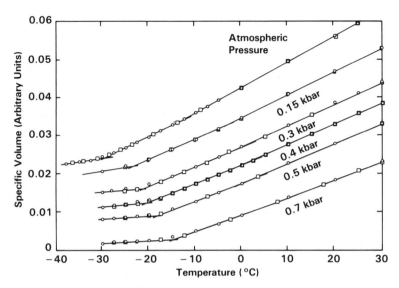

Figure 6.7   Volume isobars for poly(propylene): ○, cooling; □, heating. Reprinted from *Journal of Research of the National Bureau of Standards*; E. Passaglia and G. M. Martin, *J. Res. Natl. Bur. Stand.*, **68A**, 273 (1964).

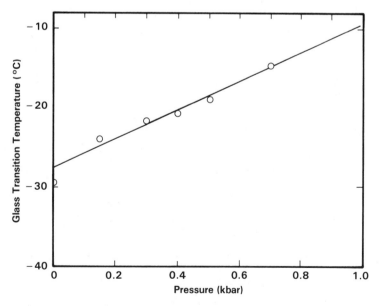

Figure 6.8   Pressure dependence of glass transition in poly(propylene). Reprinted from *Journal of Research of the National Bureau of Standards*; E. Passaglia and G. M. Martin, *J. Res. Natl. Bur. Stand.*, **68A**, 273 (1964).

The device consists of a hardened steel piston driven by the hydraulic ram of a universal test machine (such as an Instron machine) into a tightly confined cylindrical bore. Generally, for solid polymers, no sealing plug is required if the clearance between the cylinder and piston is about 5 μm. If the solid is soft with a low shear modulus, such as with poly(tetrafluoroethylene) or poly(ethylene oxide), there is a tendency for the solid to extrude between the piston and cylinder. This binds the piston, thus preventing further measurements.

The Warfield device [33], Figure 6.9, consists of two cylinders that are held together by four bolts and aligned with four guide pins. The sample also acts as a guide when the two sections are brought together. The inner bushing of both cylinders is made of hardened steel, and the inner surface of the bore is lapped and highly polished to an inside diameter of 0.635 (+0.0025, −0.0000) cm. This inner bushing is tightly fitted into an outer bushing, which is made of softer steel. Piston–cylinder devices are commonly made of steel, but nickel alloys or sintered alumina are used when high-temperature operation is desired. Beryllium copper can be used if a nonmagnetic material is required. Diamond and beryllium are used for X-ray studies. The maximum pressure that can be obtained in a standard piston–cylinder device depends on the bursting strength of the cylinder. Steel can be used up to about 20 kbar and carboloy (cobalt-bonded tungsten carbide) can be used up to 50 kbar.

It is necessary to build the device as two halves to remove the sample easily when the test is concluded. After removing the bolts and guide pins, it is usually possible to remove the sample by tapping it gently.

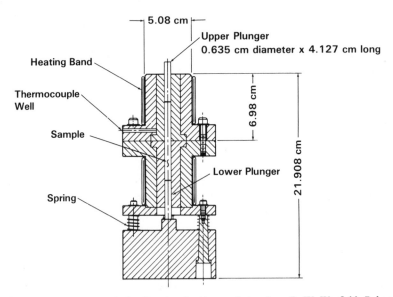

Figure 6.9 Piston–cylinder device. Reprinted with permission from R. W. Warfield, *Polym. Eng. Sci.*, **6**, 1 (1966).

Temperature measurements are made by placing heating coils around the cylinder. When the desired temperature is reached, one must wait approximately 1 h or longer for each 10°C temperature change to ensure thermal equilibrium. Two thermocouple wells are located in the inner bushing near the sample.

In preparation for an experimental run, a sample 7.62 cm long and 0.635 ($+0.0000$, -0.0025) cm in diameter is placed in the bore of the device, and the two halves are bolted together. Note, the need for an accurately machined or molded sample is a limitation of the piston–cylinder method. Once the sample is in position, two case-hardened steel plungers 4.128 cm long and 0.635 cm in diameter are inserted in the open ends of the bore above and below the sample. The device is mounted on a base, and the entire assembly is placed in a testing machine. The sample is compressed by pressing down on the steel plungers. The method used to drive the piston into the cylinder depends on the force needed to produce the desired pressure. Low values can be achieved by manually operating a lever. Higher values are attained by a hydraulic press.

By measuring the plunger travel and the load applied, we obtain a continuous stress–strain plot, in contrast to the dilatometer, in which discrete $P - V$ points are usually obtained. Linear strain is defined as $\Delta L/L_0$, where $\Delta L = L_0 - L$, L_0 is the initial length, and L is the length under load. Since the cross-sectional area does not change during the test, the linear strain is assumed to be equal to the volume strain $\Delta V/V_0$. If a frictionless piston of cross-sectional area A is pushed by a force F into a cylinder containing a fluid, it produces a hydrostatic pressure equal to F/A. For a solid this is only an approximation. The piston–cylinder device can be used as an absolute measurement; but, especially for high-pressure work, published values of transitions should be checked.

The rate of loading chosen must be slow enough to approximate equilibrium conditions. Experience indicates that rates of loading corresponding to strain rates of less than $0.008\,\text{min}^{-1}$ approximate equilibrium in polymer solids, whereas more rapid loading does not. Ideally, one should conduct an experiment by imposing a stepwise reduction in volume on the sample and measuring the accompanying pressure as a function of time, although this procedure is not common.

The precision of these measurements was considered and it was observed that duplicate measurements of volume on poly(styrene) agree within 1% [4]. The absolute accuracy is more difficult to assess, but it is estimated to be 2%. Some of the experimental error is caused by the volume changes that occur in the containing vessel, pistons, loading apparatus, and pressure-transmitting medium, all of which are neglected. Also, some of the applied pressure may be used to overcome frictional forces that can occur between the polymer sample and the wall of the bore. To check friction effects, experiments were conducted on two samples of different length [34]. Poly(styrene) of the standard length of 7.62 cm was measured as well as a sample 0.635 cm long, giving a length-to-diameter ratio varying from 1 to 12. Both stress–strain plots were identical in shape. It was concluded that friction is not a significant source of error. Similar results were obtained with various other polymers.

As mentioned earlier, the limit of a standard piston–cylinder device is about 50 kbar. Various modifications were used to raise this value. A compound cylinder can raise the bursting strength if the cylinder is subjected to an external pressure, which helps to balance the internal pressure. This is accomplished by shrinking a second cylinder onto the first so that the inner cylinder is in compression and the outer one is in tension.

Hall [39] developed a variation of the piston–cylinder device called the *belt apparatus* that is capable of reaching 150 kbar with temperatures as high as 2000°C. The device consists of a supported conical vessel where the solid sample is placed. The sample is compressed between two conical pistons that are driven by a hydraulic press. The pistons and the containing vessel have steep sides of varying slope; since the vessel is convex, contact is initially made with the middle of the conical face of the piston. Pressure from the pistons is transmitted to the sample in the containing vessel by pyrophyllite, a hydrous aluminum silicate that also serves as electrical and thermal insulation. As the piston advances, the pyrophyllite is compressed and partially extruded. The sample can be heated electrically if necessary. The supporting rings around the sample chamber form a torus around the sample vessel and give this device its name.

3.2   Representative Results

Figure 6.10 is an example of the raw data from a piston–cylinder device. The load versus length change at various temperatures for poly(propylene) is shown [40]. The load is converted to stress by dividing by the sample cross-sectional area, and the length change is converted to strain by dividing by the initial sample length. The change in shape of the 154°C isotherm compared with the other isotherms is a result of the start of the melting transition. The blip about one-half way up the 154°C isotherm results from instrument jitter, and it is not a material property.

As an example of the use of a piston–cylinder device to determine phase transitions, the crystal–crystal phase transitions in poly(tetrafluoroethylene) are shown in Figure 6.11, along with the melting phase line [41]. Three different crystal forms exist in the solid phase. The transitions between these phases are very sluggish and somewhat diffuse.

It was assumed to this point that the measurements are made on a macroscopically isotropic solid, either because the solid is amorphous or because it is polycrystalline with random orientation of the crystallites. Measurements were also made on anisotropic solids in a piston–cylinder apparatus that allows X-ray diffraction patterns to be determined as a function of pressure. Measurements in nylon-6 that was oriented by stretching show the polymer morphology as molecular chains of covalent bonds in the fiber axis direction with intermolecular hydrogen bonds bridging the chains. These sheets are then stacked up and held together by weak van der Waals bonds. The bulk modulus along the chain direction is 387 kbar, but it is only 95 kbar between planes [42]. Note that the modulus in the chain direction is close to that for metals. The relatively low overall bulk modulus of polymers is a result of their large free volume and weaker bonding between chains.

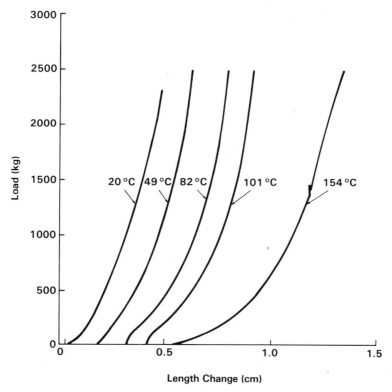

Figure 6.10 Piston–cylinder load versus length change at various temperatures for poly(propylene). Reprinted with permission from R. W. Warfield, unpublished results.

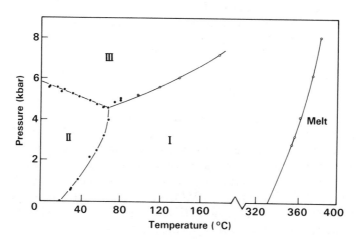

Figure 6.11 Phase diagram of poly(tetrafluoroethylene). Reprinted from C. W. F. T. Pistorius, *Polymer*, **5**, 315 (1964), by permission of the publishers, Butterworth & Co. (Publishers) Ltd. ©.

4   ACOUSTICS

4.1   Description of Apparatus

An alternate method of determining bulk modulus is to measure the sound speed, since sound speeds are related to modulus values by the relations [20]

$$v_L = [(K + 4G/3)/\rho]^{1/2} \tag{20}$$

$$v_G = (G/\rho)^{1/2} \tag{21}$$

where v_L is the longitudinal sound speed, v_G is the shear sound speed, and ρ is the density. These are the only two types of acoustic waves that can be propagated in an unbounded isotropic solid. In a longitudinal wave, the material particles vibrate along the direction of propagation. In a shear wave, the particle motion is perpendicular to the direction of propagation. Longitudinal waves are also called *dilatational, compressional,* or *irrotational waves.* Shear waves are also called *distortional, isovoluminous,* or *transverse waves.*

Because pressure variations occur rapidly in a sound wave, heat does not have time to flow in and out of the system, and the bulk modulus value determined in an acoustic measurement is an adiabatic (constant entropy) bulk modulus K_S in contrast to the isothermal bulk modulus obtained from static measurements of volume. The two values are related by the standard thermodynamic formulas

$$K_S/K_T = C_P/C_V = 1 + (TV\alpha^2/C_P) \tag{22}$$

where C_P and C_V are specific heats at constant pressure and volume, respectively.

The difference between the bulk moduli in (22) is related to the volume change that accompanies the isothermal process but does not occur in the adiabatic process. The difference amounts to a few percent for metals but can be 20% for polymers [43]. For the shear modulus, there is no difference between the adiabatic and isothermal moduli

$$G_S = G_T \tag{23}$$

since there is no volume change in shear deformation.

In addition to the adiabatic–isothermal difference, an acoustically determined bulk modulus is higher than one determined from static volume measurements because of the frequency dependence of the modulus. This effect is minimal in an elastic material, such as a metal, but it can be significant in a viscoelastic solid. For example, in PMMA at room temperature and 1 bar [6], the bulk modulus is 58.75 kbar at 6 MHz and 59.14 kbar at 30 MHz. Another possible difference between acoustic measurements and static measurements is strain dependence. Acoustic measurements are independent of strain since the strain is so small [44], only about 10^{-6}–10^{-9}. Static measurements (i.e., dilatometer or piston–cylinder) involve strains of 10^{-4} or more and there may be strain dependence.

Figure 6.12 illustrates the immersion technique [3, 45], which is a common acoustic technique for determining bulk modulus. In this method, acoustic waves are generated by a piezoelectric transducer. When this transducer is subjected to an oscillating electric field, its thickness will alternately increase and decrease with the same frequency as the electric field. Similarly, when a mechanical oscillation is applied to the transducer, it produces an oscillating electric field. Thus, transducers are used both to generate and to detect acoustic waves. Depending on its use, a transducer is called a *transmitter* or a *receiver*. Common transducer materials are quartz and various polycrystalline ceramics, such as lead zirconate titanate (PZT), polarized in a strong electrostatic field.

Measurements are made by sending acoustic pulses, typically less than 1-ms duration, through the sample. Pulses, rather than continuous waves, are used because it is easy to determine the time it takes for a pulse to go through a sample by looking for the beginning of the pulse. In the immersion technique, the sample, transmitter, and receiver are all immersed in a liquid. Pulses are sent from one transducer to the other both with and without the sample in the path of the sound beam. From the change in the transit time that occurs when the sample is removed from the path of the sound beam, the sound speed can be determined.

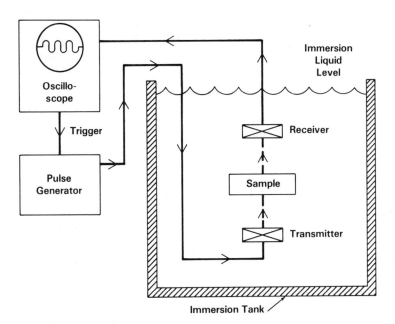

Figure 6.12 Schematic immersion apparatus.

When a sample of lateral dimensions several times the acoustic wavelength is supported with the face perpendicular to the path of the sound beam, longitudinal waves are generated in the sample. In a variation of this method, the sample is held at an angle to the sound beam, generating shear waves in the sample.

Four advantages of the immersion technique are (1) the intimate contact of the liquid with the transducers and sample provides good transfer of acoustic energy (good coupling) between the transducer and the sample, (2) the coupling is very reproducible, (3) shear waves can be generated with the same transducer that generates longitudinal waves, and (4) multiple samples can be quickly moved into and out of position.

To make shear measurements, the sample can be rotated with respect to the sound beam or, as in the apparatus described here, the transducers are rotated [3]. Eight samples are mounted on the periphery of a circle, and they are moved into and out of the path of the sound beam. Especially when making measurements as a function of temperature, there is a considerable savings of time when eight samples are immersed at once.

Lead zirconate titanate transducers of 2.5 cm diameter are used. The larger the diameter at a given frequency, the better collimated is the acoustic beam. At the usual operating frequency of 2 MHz, diffraction attenuation is negligible and sample alignment is not critical. The resonant frequency of the transducer is proportional to its thickness, 0.25 cm, and is 0.75 MHz. The maximum output of the transducer is at its resonant frequency, but it can be operated below resonance without much loss; it can also be operated at harmonics of the resonant frequency to obtain higher frequency measurements. In this manner, the apparatus is used to make measurements over the frequency range of 0.1–10 MHz, although most of the results are at 2 MHz. The central wire of a miniature coaxial cable is soldered to the back of the transducer for one electrical lead. The ground lead is that part of the device that presses against the front face of the transducers. The brass transducer holders are open to the atmosphere, thus providing air backing for the transducers, which improves their operation.

An ideal immersion liquid has a low sound speed, is liquid over a wide temperature range, and is safe to handle (alcohol, e.g., is a fire hazard). A low-viscosity silicone fluid was chosen for this apparatus. This liquid can be used over the temperature range from -50 to $150°C$.

In the simplest case, only two electronic components are needed to generate and detect the electrical oscillations: a pulsed oscillator and an oscilloscope. Pulse lengths of 10–20 cycles are used, which at 2 MHz gives a pulse length of 5–10 μs. The Fourier transform of such a pulse does not contain a large high-frequency component. Therefore, the group velocity is equal to the phase velocity of a continuous wave of the same frequency. To avoid interference, the pulse length should be less than the distance between transducers. In this apparatus, the transducers are separated by 20 cm, and the pulse length is about 2 cm. The output of the receiver is displayed on an oscilloscope; the sweep is triggered by the start of the transmitted pulse. The pulse is repeated at a rate of 60 times per second.

Longitudinal speed measurements are made with the sample held perpendicular to the path of the sound beam. The detected pulse is displayed on the oscilloscope, and the position of the first peak of the pulse is noted. The sample is then removed from the path of the sound beam. Since the sound speed in the immersion liquid is less than the sound speed in the solid sample, the signal takes a longer time to reach the receiver, and the detected pulse moves to the right on the oscilloscope screen when the sample is removed. The difference in transit times t for the sample and an equal thickness of liquid is determined from this shift in position. For a sample of thickness L, the transit time is L/v_L, where v_L is the longitudinal speed in the sample. The transit time through an equal thickness of liquid is $L/v(\text{liq})$, where $v(\text{liq})$ is the speed in the immersion liquid. Therefore,

$$\Delta t = L/v(\text{liq}) - L/v_L \qquad (24)$$

so that v_L can be calculated using the measured [3] sound speed in DC200 silicone liquid of 2-cm^2/s room temperature viscosity

$$v(\text{liq}) = 976 - 2.5(T - 25) \qquad (25)$$

for $v(\text{liq})$ in meters per second and T in degrees Celsius. For a PMMA sample of 1.27 cm thickness at room temperature, $\Delta t = 8.3\,\mu s$ and $v_L = 2690\,\text{m/s}$. Using a standard oscilloscope, we obtain about 1% accuracy of the measurements, but it can be improved to about 0.1% by the addition of a time-mark generator.

Shear speed is measured by rotating the transducers so that the sound beam strikes the sample at an angle. At off-normal incidence, both shear and longitudinal waves are generated in the sample. When both waves are present, they overlap in the received signal and are difficult to separate. However, if $v(\text{liq})$ is less than v_L, there is a critical angle beyond which there is total internal reflection of the longitudinal wave, and only the shear wave propagates through the sample. Hence, a low sound speed immersion liquid is desirable. [Note that the shear wave does not propagate through the immersion liquid. At the far end of the sample, the shear wave is converted to the only type of propagation possible in a liquid, with speed $v(\text{liq})$]. Shear speed v_S is calculated in a manner similar to that for v_L. The procedure is more involved only because the path length through the sample is somewhat greater than the sample thickness, depending on the angle at which the sample is held

$$v_S = v(\text{liq})\{[\cos\vartheta - v(\text{liq})\Delta t/L]^2 + \sin^2\vartheta\}^{-1/2} \qquad (26)$$

For a 1.27-cm-thick PMMA sample at room temperature, $\vartheta = 30°$, $\Delta t = 4.4\,\mu s$, and $v_S = 1340\,\text{m/s}$.

Other techniques for making sound speed measurements in solids utilize the delay rod technique, where the acoustic path of liquid between the transducers and the sample is replaced with solid rods, usually of quartz or metal. The purpose of the delay rods (also called *buffer rods*) is to give enough time between pulses so that the complete pulse can be sent by the transmitter before the beginning of the pulse arrives at the receiver. This technique eliminates stray radio frequency pickup by the receiver. A very thick sample could be used for the same purpose, but, especially for polymers, the absorption may be so high that detection of the signal is difficult.

In the delay rod technique, longitudinal and shear measurements are made separately with different sets of transducers. Quartz transducers are often used because quartz crystals produce different vibrations depending on how they are cut. An X-cut quartz crystal is used to generate longitudinal waves, while Y-cut or AC-cut crystals are used to generate shear waves.

Good bonding between the transducers and the delay rods and between the delay rods and the sample is important. For longitudional waves, silicone liquid, stopcock grease, and glycerin were used successfully. For shear waves, which are more highly damped, the problem is more critical. A low molecular weight poly(α-methylstyrene) liquid and mixtures of phthalic anhydride and glycerol are effective and are easily removed. Epoxy adhesives give good bonding, but they are difficult to remove.

Sound speed can be measured in the same manner as the immersion technique or by using a null method. In the null method, the pulse input to the transducer is split in two . The pulse that has traversed the sample is then mixed with an out-of-phase signal directly from the pulsed oscillator. By properly delaying and attenuating the direct signal, a null is obtained. Using this technique, we obtain an accuracy of 0.5%. Higher accuracy is achieved with better electronics. Variations of this technique [6, 8, 45] are useful to accuracies of 0.05%.

4.2   Representative Results

A table of longitudinal and shear sound speeds for selected solids is given in Table 6.2, which was compiled from various sources [3, 46, 47]. Density is also listed since it is needed to calculate bulk modulus from sound speeds and because acoustic properties are sensitive to density. Two liquids are also listed for comparison purposes. As expected, diamond has the highest sound speeds and metals have higher speeds than polymers (with the exception of lead). The sound speeds in Table 6.2 vary by a factor of 10 compared with the bulk modulus values in Table 6.1, which vary by a factor of 400.

Like bulk modulus, sound speeds decrease as temperature increases. As an example, Figure 6.13 shows the temperature dependence of sound speeds for poly(propylene) [3]. The temperature dependence is rather high in this case, -15 m/s K for the longitudinal speed and -6.7 m/s K for the shear speed. For PMMA, the corresponding values [3] are -2.5 and -2.0 m/s K. Sound speeds

Table 6.2 Typical Acoustic Properties[a]

Material	Density (g/cm^3)	Longitudinal Speed (m/s)	Shear Speed (m/s)
Diamond	3.512	18299	12804
Aluminum	2.70	6420	3040
Iron	7.85	5960	3240
Quartz	2.2	5900	3750
Tungsten	19.3	5220	2890
Copper	8.93	4760	2325
Platinum	21.4	3260	1730
Epoxy polymer	1.205	2820	1230
Nylon-6,6	1.147	2710	1120
Nylon-6	1.146	2700	1120
Poly(methyl methacrylate)	1.191	2690	1340
Poly(propylene)	0.913	2650	1300
Poly(oxymethylene)	1.425	2440	1000
Poly(ethylene), high density	0.957	2430	950
Poly(styrene)	1.052	2400	1150
Lead	11.4	2160	700
Poly(vinylidene fluoride)	1.779	1930	775
Water	0.998	1497	
Mercury	13.5	1450	

[a]Data taken from [3, 46, 47].

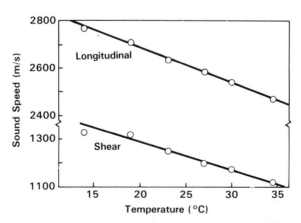

Figure 6.13 Temperature dependence of sound speeds for poly(propylene). Reprinted with permission from B. Hartmann and J. Jarzynski, *J. Acoust. Soc. Am.*, **56**, 1469 (1974).

increase as pressure increases. For PMMA [6], the pressure derivative of the longitudinal sound speed is 0.266 m/s bar; for shear speed it is 0.107 m/s bar.

Reference [48] summarizes the relations between polymer properties and sound speeds, as well as to bulk modulus. In addition to density, properties considered include the state of cure, a variety of transitions, and composite materials.

5   DIAMOND ANVIL CELL

5.1   Description of Apparatus

In the diamond anvil cell (DAC) the sample is placed between the flat parallel faces of two opposed diamond anvils and subjected to pressure when a force pushes the two anvils together. The basic configuration [49] is shown in Figure 6.14. The device is small enough to fit in the palm of the hand, inexpensive, and easy to operate and maintain. Pressures of 1 Mbar or higher can be generated, although the accuracy of the pressure calibration is questionable above 300 kbar. The high pressure can be combined with high temperature using laser heating. In general, the accuracy of volume measurements is not as good as for a dilatometer or piston–cylinder device. An additional limitation of the DAC is the small size of the sample, typically 0.5 mm in diameter and 0.1 mm thick. Diamond is used because it is the highest modulus substance known (see Table 6.1) and because it is transparent to X-rays and light, which allow a variety of

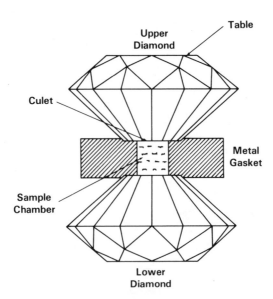

Figure 6.14   Basic diamond anvil cell. Reprinted with permission from A. Jayaraman, *Rev. Mod. Phys.*, **55**, 65 (1983).

measurements to be made under pressure. This presentation is adapted from the review of Jayaraman [49].

There are several implementations of the basic configuration shown in Figure 6.14. They vary in the way the force is generated and how the anvil alignment is maintained. In the version used at the National Institute of Standards and Technology [50], force is applied using Belleville spring washers attached to a lever arm with a 2:1 mechanical advantage (see Figure 6.15). The main body is a rectangular plate $15 \times 6 \times 2$ cm made of American Iron and Steel Institute (AISI) 4340 or Vascomax 300 steel. Force is produced by compressing the Belleville spring washers by turns of the screw. The force on the pressure plate is transmitted to a movable piston containing a hemispherical mount for the lower diamond. The mount can be tilted in its socket by adjusting screws to secure parallel alignment of the diamond anvil flats, as determined by an optical interference pattern. The stationary diamond mount can be translated for centering purposes by three, symmetrically placed adjusting screws; once centered it is locked in position by two pulling screws. The diamond mounts are hardened to 55–60 on the Rockwell C scale. The mounts have apertures for entry and exit of optical, X-ray, and other radiations through the diamonds.

Diamonds for the anvils are usually selected from brilliant-cut gemstones that vary from $\frac{1}{8}$ to $\frac{1}{2}$ carat. The culet (base) of the diamond is removed by grinding a flat octagonal surface, whose area is typically 0.45 mm^2 and 0.7 mm from side to side of the octagon. The octagonal surface that is opposite the anvil flat, referred to as the table, has a diagonal distance from 2 to 4 mm, depending on the size of the diamond. In general, larger diamonds are preferred for higher pressure and for larger pressurized volume.

A metal foil gasket is used to contain a pressure-transmitting medium. The initial thickness of the foil is about 0.2 mm. The foil is indented by being pressed between the diamond anvils to a thickness of about 0.1 mm. In the center of this

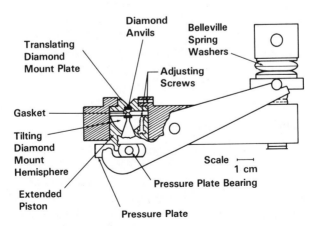

Figure 6.15 National Institute of Standards and Technology diamond anvil cell. Reprinted with permission from A. Jayaraman, *Rev. Mod. Phys.*, **55**, 65 (1983).

indentation a 200-μm-diameter hole is drilled. The gasket is then replaced onto the lower diamond flat, and the cell is ready to be loaded. Among others, tempered 301 stainless steel, Inconel X750, and Waspalloy gasket materials are used.

The DAC was first introduced in 1959, but it became widely used only after the ruby fluorescence method of pressure calibration was developed in the 1970s. A tiny chip of ruby (chromium-doped alumina) 5–10 μm in size is placed in the DAC along with the sample. Fluorescence is excited by either a He-Cd laser line or any source of strong light. The R lines of ruby are quite intense, and the doublets R_1 and R_2 have the wavelengths 6927 and 6942 Å, respectively, at atmospheric pressure. Under pressure, they shift to higher wavelengths. Using well-established freezing points of several liquids and some solid–solid transitions as fixed points, it was shown that the R lines shift linearly with pressure at the rate of 0.365 Å/kbar up to 300 kbar. There are some questions about the exact values of the nonlinear behavior above this pressure, but ruby fluorescence remains the best method of determining pressure up to 1 Mbar.

Several pressure-transmitting media were used in the DAC. Methanol–water mixtures are useful up to about 200 kbar, and liquid helium and hydrogen are nearly hydrostatic to above 600 kbar, although they require cryogenic or high-pressure filling.

Various techniques are used to monitor sample changes produced by the pressure generated in the DAC. Included are X-ray diffraction, Raman scattering, Brillouin scattering, electrical resistance, optical absorption, and optical reflectivity. These techniques can be used to detect phase changes. X-ray measurements can also be directly related to change in crystal volume with pressure and hence to modulus.

Brillouin scattering allows the measurement of sound speeds as a function of pressure and the bulk modulus can be calculated once the density is known. In Brillouin scattering, a laser-generated hypersonic wave (i.e., a very high frequency sound wave) traveling through a crystal produces a periodic modulation of the refractive index in the form of planes of higher and lower refractive index. Light can be scattered from these planes in a manner analogous to Bragg's law. However, since the planes are nonstationary, the scattered light undergoes a Doppler shift in frequency, depending on the velocity of the hypersonic wave and the geometry of interaction of the light.

5.2   Representative Results

Of the solids listed in Table 6.1, cesium has the lowest bulk modulus. The exact value for cesium is somewhat in question because the data are very nonlinear, but it is certain that the bulk modulus is low. In addition, cesium undergoes an electronic transition at relatively low pressure. It is believed [49] that cesium undergoes a 6s–5d electronic transition at 42 kbar. All the anomalous physical properties of cesium under pressure are attributed to this transition.

At high enough pressure, it is theorized that all electrical insulators will become metallic conductors. Of particular interest is the quest for metallic hydrogen. From these studies it was discovered that hydrogen is an excellent hydrostatic pressure medium up to several hundred kilobars even though it is solid at these pressures. Hydrogen solidifies near 55 kbar at 298 K to a clear transparent solid, and it remains that way up to at least 500 kbar. Using the DAC, Shimizu and co-workers [51] determined an analytic equation of state for solid hydrogen at room temperature from which it follows that the bulk modulus is

$$K = 40.9\rho P^{2/3} \tag{27}$$

where P and K are expressed in kilobars and ρ is in grams per cubic centimeter. The equation is valid in the pressure range of 55–200 kbar. At 100 kbar, $K = 280$ kbar, which is comparable to the bulk modulus of liquid mercury at 1 bar. Thus, solid hydrogen is a soft solid.

Iodine is a molecular crystal that undergoes a continuous pressure-induced transformation to the metallic state in a region easily reached by the DAC. Iodine crystallizes in an orthorhombic space group. It retains this structure and remains a molecular crystal up to 206 kbar, even though it becomes metallic above 170 kbar. Metallization occurs by conduction–valence-band overlap prior to dissociation [49].

High-pressure X-ray diffraction studies on praseodymium to 300 kbar in the DAC showed [52] that there is a phase transition from distorted face-centered cubic (fcc) to distorted hexagonal close-packed (hcp) at 210 kbar. A 19% volume change is observed, as shown in Figure 6.16. This is interpreted in terms of a change in the valence state from the 3+ toward the 4+ state. There is a tendency for praseodymium to be quadrivalent judged from its chemical properties; for example, PrO_2 exists.

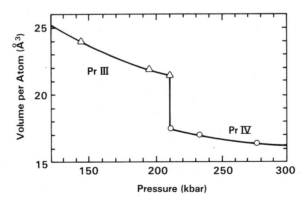

Figure 6.16 Volume isotherms for praseodymium. Reprinted with permission from H. K. Mao, R. M. Hazen, P. M. Bell, and J. Wittig, *J. Appl. Phys.*, **52**, 4572 (1981).

6   SHOCK WAVES

6.1   Description of Apparatus

The use of shock waves as a way to study the high-pressure properties of solids is a specialized and extensive area of research. This section gives a flavor for the field that is based on the review by Duvall and Graham [53]. Shock waves are nearly discontinuous jumps in pressure traveling faster than the speed of sound and can generate the highest reported pressures, as high as 30 Mbar, near an underground nuclear explosion [54]. Various techniques are used in the laboratory to generate pressures up to several megabars. In the contact explosive method, strong one-dimensional shocks are produced by lenslike combinations of fast and slow detonating explosives that make it possible to convert detonation initiation at a point to an essentially plane detonation wave up to about 30 cm in diameter and flat to within 1 or 2 mm. The diameter-to-thickness ratio must be large to avoid lateral release waves from the boundaries, which is a requirement to approximate uniaxial loading. The simultaneity of the wave is about 10 ns, and the total duration of the experiment is typically 1 μs. In this manner, 100–400-kbar pressures are generated routinely in aluminum. (The magnitude of the pressure depends not only on the explosive used, but also on the solid being tested.)

Shock-wave compressions are designed to be one-dimensional, but the stresses are often much greater than the yield strength of the solid so that shear effects are usually neglected and the sample is treated as a perfect fluid. It is assumed that the compression for a given stress is the same as that which would be produced by a hydrostatic pressure of the same magnitude. This assumption was tested widely and no significant discrepancies are indicated between shock and static loading measurements.

Another technique used to generate shock waves is the flyer plate method in which a 1.5-mm-thick steel plate is accelerated, either explosively or by using compressed gas, over a free run of a few centimeters before impacting the sample. Typical impact velocities range from 1 to 7 km/s. Pressures of 100–1000 kbar in aluminum and of several megabars in iron were generated. A variation of this technique allows one to separate the bulk and shear effects by impacting a flyer plate onto a sample at an angle to the direction of travel. Results were presented [55] for a polyurethane polymer up to 14 kbar, which is a low-pressure region where the usual neglect of shear modulus is a questionable assumption.

In the projectile-impact technique, a flat-faced impactor is accelerated, in vacuum, in a smooth bore gun barrel. Acceleration can be effected either by compressed gas or a propellant. Bore diameters of 6–10 cm are common. Barrel lengths vary from 3 to 24 m. Simple gas gun velocities range from 30 cm/s to 1.5 km/s. Up to 8 km/s can be reached with a two-stage light gas gun. Alignment of the sample relative to the impactor is critical. Tilt values of 500 μrad are normally acceptable for a projectile velocity of 1 km/s.

A particular interest in the use of shock waves is the study of shock-induced polymorphic transitions. Questions have been raised about the validity of such results considering the short time available for the transition to occur. Under static loading, transformation rates are dominated by the statistical probability of forming nuclei in a low-defect solid [53]. Under shock loading, the plastic deformation required to achieve the high pressure is a consequence of the creation and motion of copious quantities of defects, chiefly dislocations, and the reaction rate is increased by orders of magnitude.

6.2   Representative Results

The most widely studied shock-induced phase transition is the 130-kbar transition in iron. The transition was first discovered using shock waves and only later confirmed using static measurements. The transition is now used as a static pressure calibration point. The transition is from the α phase [body-centered cubic (bcc)] to the ε phase (hcp), and it involves a 5.4% volume change at the transition. The phase diagram for iron [53], which was compiled from various sources, is shown in Figure 6.17. The $\alpha \leftrightarrows \varepsilon$ transition at 130 kbar is a triple point with the γ phase (fcc).

Other transitions of interest include the transformation of graphite to diamond at a shock pressure of about 400 kbar with a mixed phase region extending to 600 kbar. The commercial production of industrial quality diamonds by a shock process yields crystallites ranging from 500 Å to 30 μm in size. Since diamonds are observed in meteorites, it is evident that the meteorites were

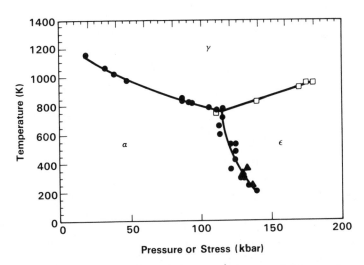

Figure 6.17   Phase diagram of iron. The symbols represent data recorded from various sources. Reprinted with permission from G. E. Duvall and R. A. Graham, *Rev. Mod. Phys.*, **49**, 523 (1977).

subjected to high pressure. Cubic and wurtzite forms of boron nitride were formed by shock compression of hexagonal boron nitride [56] and are stable at atmospheric pressure.

An extensive summary of shock-induced transitions [53] indicates that the highest pressure transitions reported are for α-Fe_2O_3 crystal, listed at 800 kbar, and for iodine at 700 kbar. These values are tentative, however, since they were not confirmed in independent tests.

7   SUMMARY

Five experimental techniques for determining the effects of pressure on solids were examined. These techniques differ in the pressure range covered and the accuracy of the measurements. The measurements are used to calculate the bulk modulus and to establish the phase diagram of solids.

A dilatometer can be used for measurements up to about 2 kbar. Isothermal bulk modulus measurements, obtained by calculating the derivative of the volume versus pressure data, are accurate to about 0.5%. The measurement is truly hydrostatic, and irregularly shaped samples can be used. All other techniques require at least two faces of the sample to be reasonably flat.

The standard piston–cylinder device can be used up to 50 kbar, while a variation of the device, known as the belt apparatus, can go up to 150 kbar. The measurement is quasihydrostatic, and bulk modulus calculations are accurate to about 2%.

Acoustic measurements are generally performed only at atmospheric pressure. Higher accuracy, up to 0.05%, can be achieved, partly because the data do not need to be differentiated. In comparing these results with other data, it must be remembered that acoustic measurements yield the adiabatic bulk modulus rather than the isothermal bulk modulus obtained in the previous two methods. Also, there may be some frequency dependence of the adiabatic bulk modulus, particularly for polymers.

The diamond anvil cell can generate pressure on the order of 1 Mbar. Accuracy is perhaps an order of magnitude less than the above methods, and this apparatus is used primarily to establish phase diagrams rather than to determine bulk modulus.

Shock waves can produce the highest reported pressures—30 Mbar in an underground nuclear explosion—but typically several megabars are generated in the laboratory. However, the accuracy obtained is poorer than that for the diamond anvil cell.

Representative data obtained using the above techniques show that the bulk modulus varies from 5550 kbar for diamond to 14 kbar for cesium. A variety of transitions were considered in establishing phase diagrams, from the glass transition in polymers to a crystal–crystal transition in iron to the formation of diamonds from graphite.

References

1. F. Birch, "Compressibility: Elastic Constants," in S. P. Clark, Jr., Ed., *Handbook of Physical Constants-Revised Edition*, Geological Society of America Memoir 97, Geological Society of America, Inc., New York, 1966, pp. 107–173.

2. L. A. Davis and R. B. Gordon, *J. Chem. Phys.*, **46**, 2650 (1967).

3. B. Hartmann and J. Jarzynski, *J. Acoust. Soc. Am.*, **56**, 1469 (1974).

4. R. W. Warfield, J. E. Cuevas, and F. R. Barnet, *Rheol. Acta*, **9**, 439 (1970).

5. R. W. Warfield and B. Hartmann, *J. Appl. Phys.*, **44**, 708 (1973).

6. J. R. Asay, D. L. Lamberson, and A. H. Guenther, *J. Appl. Phys.*, **40**, 1768 (1969).

7. R. W. Warfield, *Makromol. Chem.*, **175**, 3285 (1974).

8. D. L. Lamberson, J. R. Asay, and A. H. Guenther, *J. Appl. Phys.*, **43**, 976 (1972).

9. A. I. Gubanov and S. Yu. Davydov, *Sov. Phys. Solid State*, **14**, 372 (1972).

10. F. P. Bundy, H. T. Hall, H. M. Strong, and R. H. Wentorf, *Jr.*, *Nature (London)*, **176**, 51 (1955).

11. C. R. Wylie, Jr., *Advanced Engineering Mathematics*, 3rd ed., McGraw-Hill, New York, 1966, p. 136.

12. E. A. Moelwyn-Hughes, *Physical Chemistry*, 2nd ed., Pergamon, Oxford, 1961, p. 330.

13. R. E. Barker, Jr., *J. Appl. Phys.*, **38**, 4234 (1967).

14. M. G. Broadhurst and F. I. Mopsik, *J. Chem. Phys.*, **52**, 3634 (1970).

15. J. G. Curro, *J. Macromol. Sci. Revs. Macromol. Chem.*, **C11**, 321 (1974).

16. D. J. Pastine, *J. Chem. Phys.*, **49**, 3012 (1968).

17. B. Hartmann, *Acustica*, **36**, 24 (1976).

18. R. Simha and T. Somcynsky, *Macromolecules*, **2**, 342 (1969).

19. B. Hartmann and M. Haque, *J. Appl. Phys.*, **58**, 2831 (1985).

20. R. N. Thurston, "Wave Propagation in Fluids and Normal Solids," in W. P. Mason, Ed., *Physical Acoustics*, Vol. I, Part A, Academic, New York, 1964, pp. 1–110.

21. P. W. Bridgman, *Rev. Mod. Phys.*, **18**, 1 (1946).

22. N. Bekkedahl, *J. Res. Natl. Bur. Stand.*, **42**, 145 (1949).

23. T. Grindley and J. E. Lind, Jr., *J. Chem. Phys.*, **54**, 3983 (1971).

24. G. M. Martin and L. Mandelkern, *J. Appl. Phys.*, **34**, 2312 (1963).

25. A. Quach and R. Simha, *J. Appl. Phys.*, **42**, 4592 (1971).

26. P. Zoller, P. Bolli, V. Pahud, and H. Ackermann, *Rev. Sci. Instrum.*, **47**, 948 (1976).

27. W. A. Steele and W. Webb, "Compressibility of Liquids," in R. S. Bradley, Ed., *High Pressure Physics and Chemistry*, Vol. 1, Academic, New York, 1963, pp. 145–176.

28. E. Baer and J. L. Kardos, *J. Polym. Sci. Part A*, **3**, 2827 (1965).

29. R. D. Corsaro, J. Jarzynski, and C. M. Davis, Jr., *J. Appl. Phys.*, **45**, 1 (1974).

30. O. Olabisi and R. Simha, *Macromolecules*, **8**, 206 (1975).

31. E. Passaglia and G. M. Martin, *J. Res. Natl. Bur. Stand.*, **68A**, 273 (1964).

32. E. A. DiMarzio, J. H. Gibbs, P. D. Fleming, III, and I. C. Sanchez, *Macromolecules*, **9**, 763 (1976).

33. R. W. Warfield, *Polym. Eng. Sci.*, **6**, 1 (1966).

34. R. W. Warfield, *J. Appl. Chem.*, **17**, 263 (1967).

35. R. W. Warfield, *Makromol. Chem.*, **116**, 78 (1968).

36. R. W. Warfield, J. E. Cuevas, and F. R. Barnet, *J. Appl. Polym. Sci.*, **12**, 1147 (1968).

37. R. W. Warfield and F. R. Barnet, *Ang. Makromol. Chem.*, **44**, 181 (1975).

38. R. W. Warfield, "Static High-Pressure Measurements on Polymers," in R. A. Fava, Ed., *Methods of Experimental Physics*, Vol. 16C, Academic, New York, 1980, pp. 91–116.

39. H. T. Hall, *Rev. Sci. Instrum.*, **31**, 125 (1960).

40. R. W. Warfield, unpublished results.

41. C. W. F. T. Pistorius, *Polymer*, **5**, 315 (1964).

42. T. Ito, T. Hirata, and S. Fujita, *J. Polym. Sci. Polym. Phys. Ed.*, **17**, 1237 (1979).

43. R. W. Warfield, D. J. Pastine, and M. C. Petree, *Appl. Phys. Lett.*, **25**, 638 (1974).

44. B. Hartmann and J. Jarzynski, *J. Appl. Phys.*, **43**, 4304 (1972).

45. B. Hartmann, "Ultrasonic Measurements," in R. A. Fava, Ed., *Methods of Experimental Physics*, Vol. 16C, Academic, New York, 1980, pp. 59–90.

46. H. J. McSkimin and W. L. Bond, *Phys. Rev.*, **105**, 116 (1957).

47. R. C. Weast, Ed., *Handbook of Chemistry and Physics*, 68th ed., Chemical Rubber Company Press, Boca Raton, FL, 1987, pp. E-47 and F-80.

48. B. Hartmann, "Acoustic Properties," *Encyclopedia of Polymer Science and Engineering*, Vol. 1, 2nd ed., Wiley, New York, 1984, pp. 131–160.

49. A. Jayaraman, *Rev. Mod. Phys.*, **55**, 65 (1983).

50. G. J. Piermarini and S. Block, *Rev. Sci. Instrum.*, **46**, 973 (1975).

51. H. Shimizu, E. M. Brody, H. K. Mao, and P. M. Bell, *Phys. Rev. Lett.*, **47**, 128 (1981).

52. H. K. Mao, R. M. Hazen, P. M. Bell, and J. Wittig, *J. Appl. Phys.*, **52**, 4572 (1981).

53. G. E. Duvall and R. A. Graham, *Rev. Mod. Phys.*, **49**, 523 (1977).

54. L. V. Al'tshuller, B. N. Moiseev, L. V. Popov, G. V. Simakov, and R. F. Trunin, *Sov. Phys. JETP*, **27**, 420 (1968).

55. Y. M. Gupta, *Polym. Eng. Sci.*, **24**, 851 (1984).

56. N. L. Coleburn and J. W. Forbes, *J. Chem. Phys.*, **48**, 555 (1968).

INDEX